More than
140
Questions
PART
6
2020

Vibration Analysis Certification Exam Preparation Package

Certified Vibration Analyst Category I

Rotating Equipment

Ali M. Al-Shurafa

Vibration Analysis Certification Exam Preparation Package Certified Vibration Analyst Category I
ROTATING EQUIPMENT CAT I PREP I SERIES PRACTICE TESTS FOR CVA (PART 6)
ISBN: 978-1-64415-014-6

Disclaimers

Although every precaution has been taken to verify the accuracy of the information contained herein, the author and publisher assume no responsibility for any errors. No liability is assumed for damages that may result from the use of this package. The question banks in this booklet are related to the standard Body of Knowledge according to ISO standard 18436-2 Second Edition 2014.

Comments can be sent to info@prepcertify.com

Title	Vibration Analysis Certification Exam Preparation Package Certified Vibration Analyst Category I: Rotating Equipment	**Author**	Ali M. Al-Shurafa
Subtitle	CAT I PREP I PART 6	**Series**	CAT I PREP I SERIES PRACTICE TESTS FOR CVA
ISBN-10	1-64415-014-X	**ISBN-13**	978-1-64415-014-6
Publisher	Prep Certify	**Ordering**	www.prepcertify.com

Notes

TABLE OF CONTENTS

Notes

CAT I PREP I PACKAGE AND YOUR CERTIFICATION

PART 6 of CAT I PREP I SERIES is about "Equipment Knowledge" which is a prerequisite for any trustworthy machinery vibration analysis. Based on the details of the ISO requirements for vibration analysis Cat I, Equipment Knowledge is basically rotating equipment. This part of the series is also very beneficial for industrial practitioners dealing with rotating equipment during sales, operation and maintenance (on the site and at shops).

In this book, the focus is on the following types of rotating equipment: types pumps, compressors, electric motors and gearboxes. These assets constitute the majority of machines an analyst should see in the industry. More emphasis is given to motors and pumps but many of their questions are applicable to other types of rotating equipment. Figure 1 shows a basic classification of these assets.

Some questions in the Question Bank address the functions of these machines. As an analyst, it is expected to recognize what each piece of equipment does from the operator's perspective. For instance, all pumps and compressors in the plant increase the pressure (or flow rate) of fluids. They require a driver like a motor to rotate the shaft of the pump or compressor at the desired speed. However, a motor alone does not support the production of the plant because motors don't directly interface with the fluids/process. As a practitioner, you need to understand the basic operating parameters for these machines and their units. For example, do not mix between ips and psi. The first is a unit for velocity vibration while the second is for pressure. Both units can be used during the troubleshooting of a malfunctioning pump.

Because motors are the most common type of machines, a great deal of questions in the Question Bank test concepts related to motor's operating/design parameters such as speed, slip, load, line frequency and poles. It is very useful and important to know the line frequency in your country because you can make sense of many vibration frequencies during analysis.

To excel in machinery vibration analysis and troubleshooting, you need to obtain strong understanding of the construction of the common types of rotating machines and how they operate. A common approach to understand the machines at your facility is to read the equipment manuals from the OEMs. Ensure you

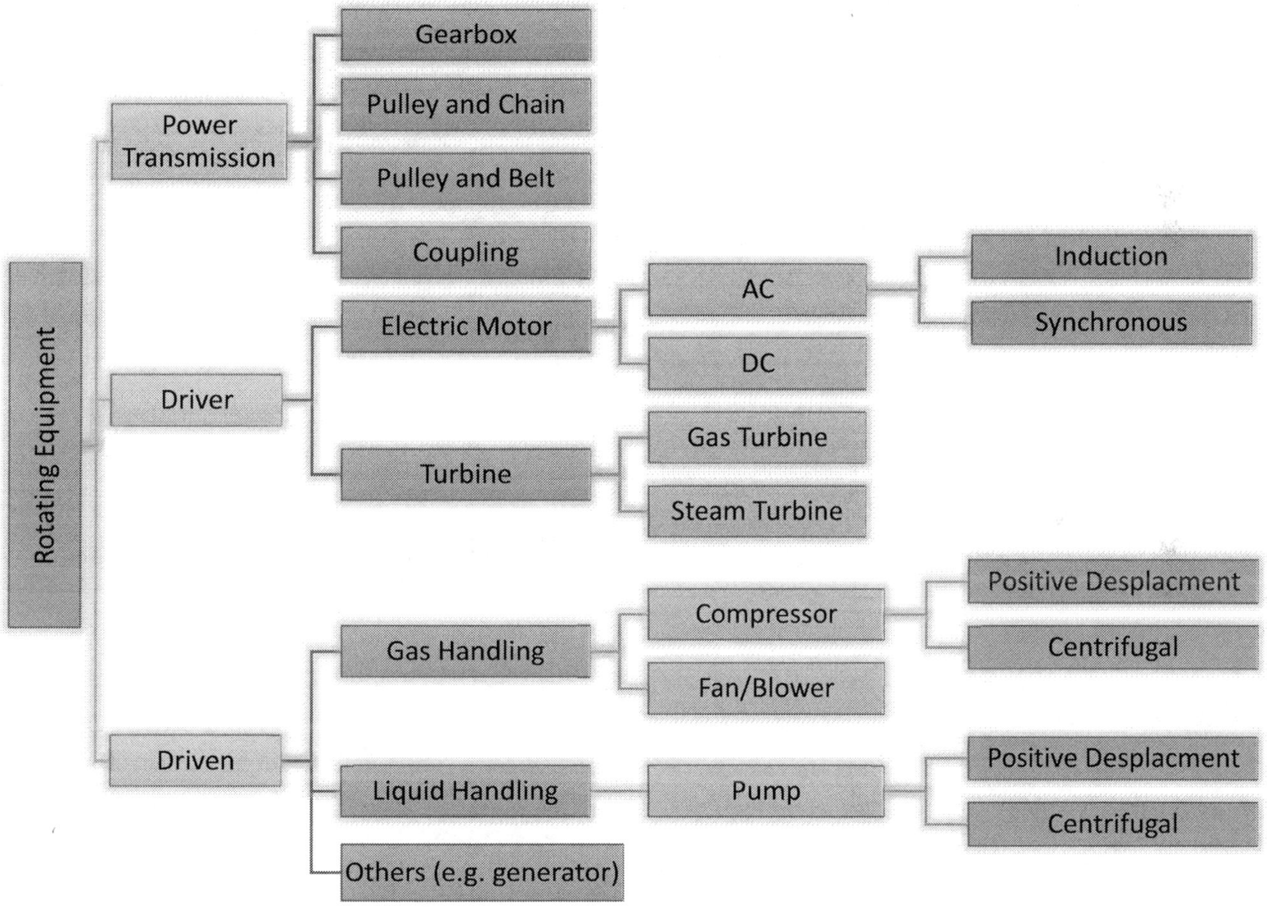

Figure 1 Common Rotating Equipment Seen in Plants

pay special attention to the cross sectional drawings and troubleshooting sections. As a fault develops, a “signal” or message is sent to the analyst from the defective part in a certain way that is related to the failure mode. Refer to Figure 2 for examples.

Another set of questions in the Bank focuses on the function of a specific part of the equipment like the impeller of a pump. This type of questions is very important for you as an analyst. It is expected to know the basic internal components to link the vibration signatures to the failure modes of these components. For example, a centrifugal pump may experience a high vibration due to cavitation whose location is the impeller. Motors do not experience vibration due to cavitation. However, both pump and motor have bearings and as a result they share common vibration problems and failure modes of bearings like looseness of a bearing in the bearing housing. Refer to Figure 2 for examples.

To link the symptoms of a defect to the faulty part, you may need to calculate the forcing frequencies. The formula sheet given in page 20 is very helpful for this step. Depending on your certification body, these formulas may not be provided to you in the exam. Most of them are simple and you are expected to use them frequently in real life analysis. So, it is very

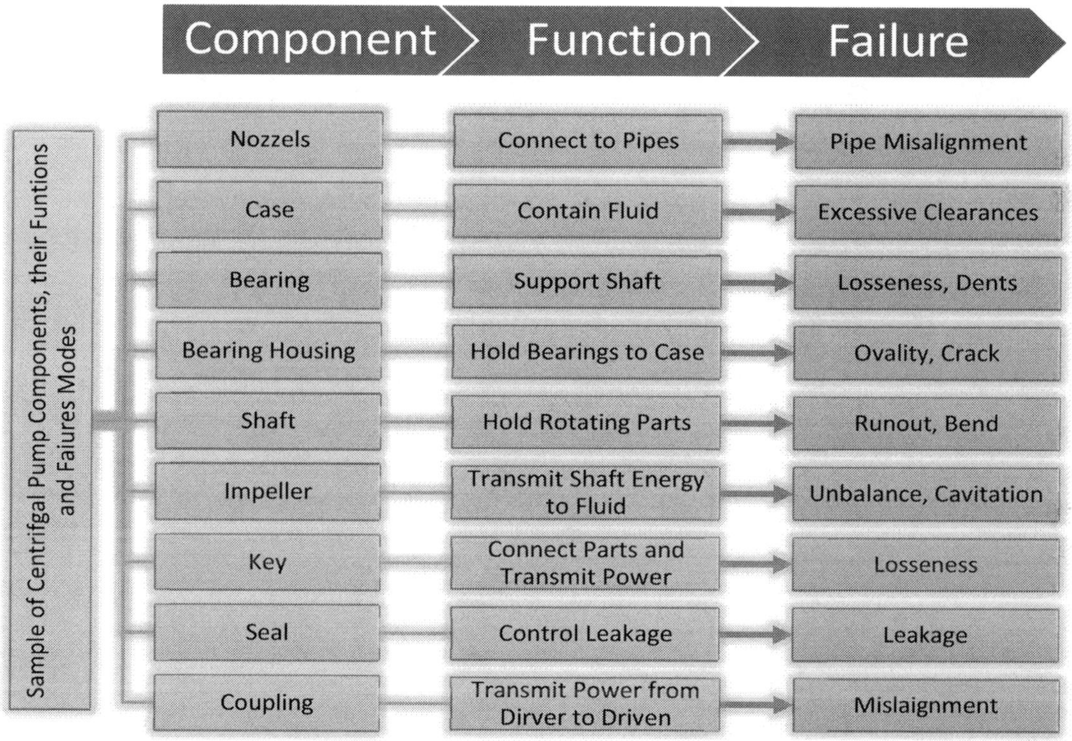

Figure 2 Simplified chart to illustrate the concept of linking pump comonents, their functions and a faiure mode example of each part. A very important concept during troublshoting.

recommended to memorize them.

To troubleshoot machinery faults, consider the following four basic steps as a guideline. These stes are explained in Figure 3.

1. Recognize the equipment type. You need to distinguish the differences between similar equipment.
2. List the key construction components in the equipment. Some components are internal while others are external and they interface with the surroundings.
3. Describe the fault based on the field observations and vibration signatures.
4. Link how each part of the machine may fail to the observation and vibration symptoms. Notice that most parts can fail in numerous ways (referred to as failure modes). For Cat I, a couple of failure modes are enough.

In practice, because of the diversity of the industry, there are synonymous terms that are necessary to be familiar with to communicate with other professionals. For instance, journal bearing, sleeve bearing and fluid film bearing are used interchangeably (despite the differences some specialists may emphasize on). Several question are included in the Question Bank to cover this aspect of the BoK.

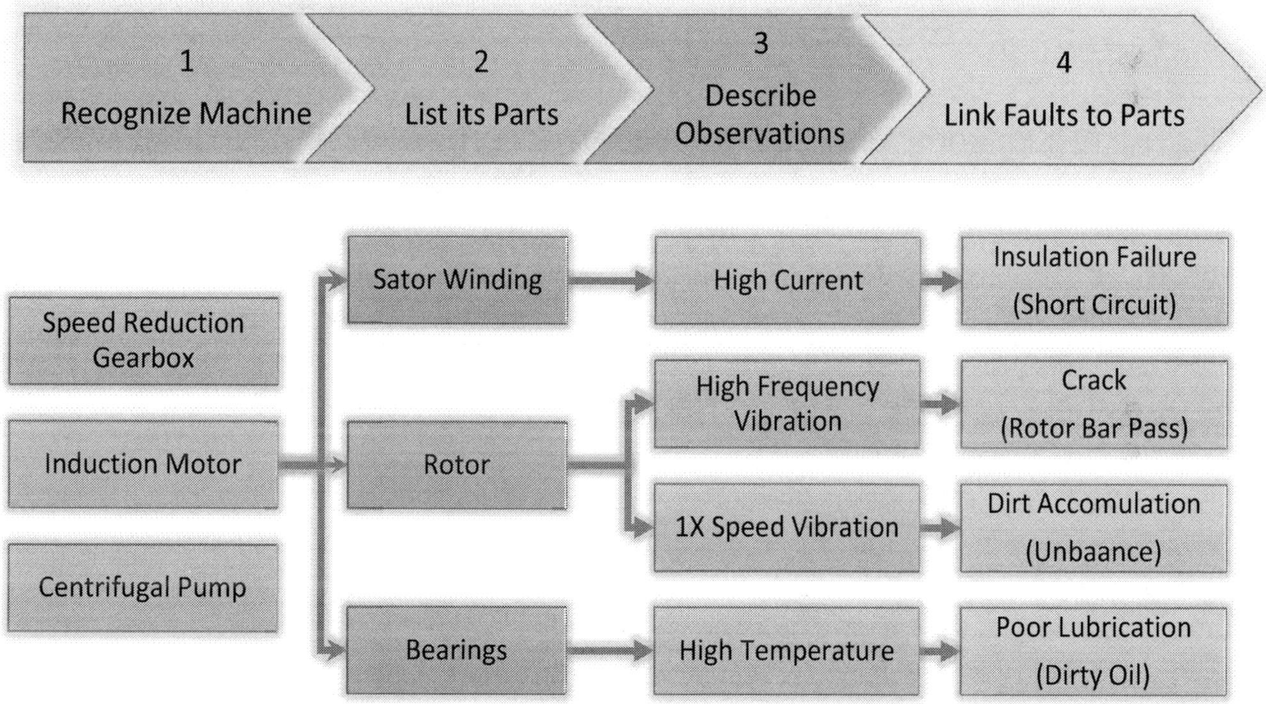

Figure 3 Four step guideline for troubelshoting rotating equipment using tyical failure modes

Reading name plates and performance curves are not clearly stated in the ISO Standard as part of the BoK. However, it's the author's opinion that these skills are foundational and very useful for vibration analysts and rotating equipment practitioners. Also, this is true for the awareness of the units used for the operating parameters in these name plates and curves. Most of these question in the Question Bank are followed by the challenge icon. The reader may notice the intentional use of many illustrations, in most of the questions. The objective of this is that you use the book as an educational tool in addition to test your preparedness for the exam.

Part 6 Question Bank includes more than 140 questions. Around 70 questions are challenging and identified with icon at the end of the question stem. This mix of difficulty makes the book is suitable for Cat I and II.

Good luck- Ali M. Al-Shurafa

CAT I BODY OF KNOWLEDGE

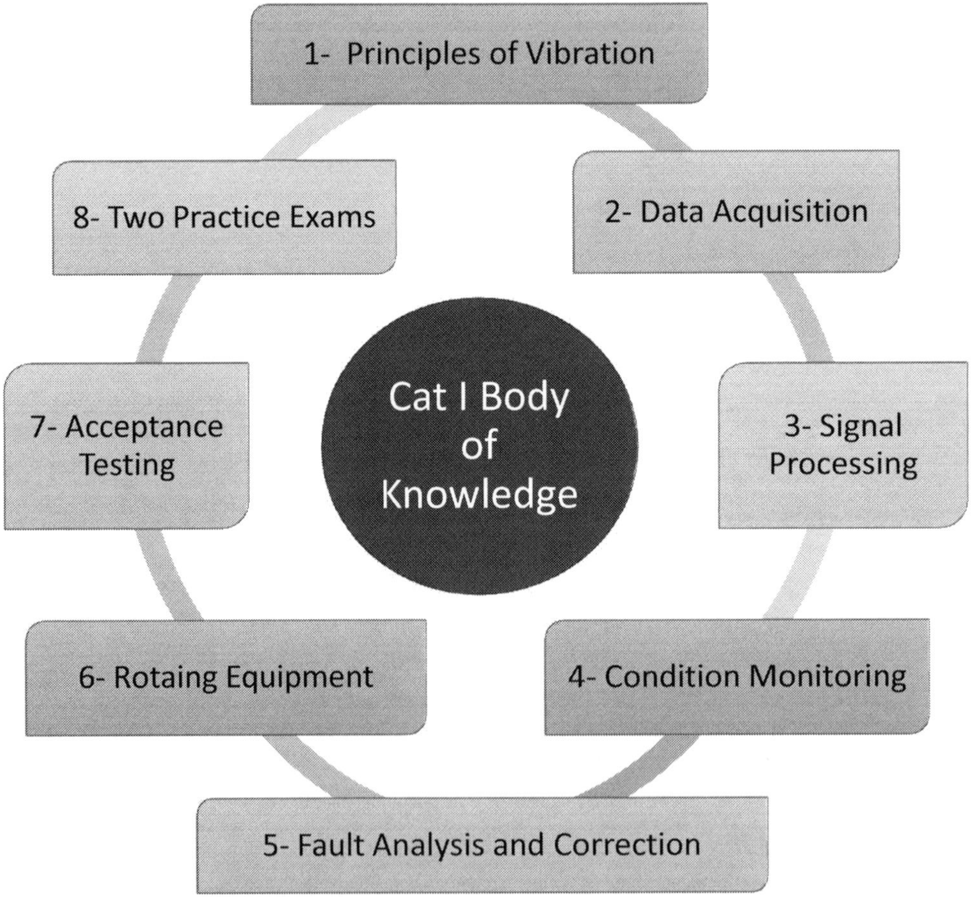

CAT I PART 6 BODY OF KNOWLEDGE

Below is the Body of Knowledge as specified by ISO 18436-2 for Part 6 "Rotating Equipment". For a complete list of topics, refer to Part 8 or the Standard.

Topic	Details
Electric motors, generators and drives	Recognize AC induction motor, and basic faults, e.g. bearing noise and damage, balance, looseness and misalignment
Pumps, fans	Recognize basic pump and fan combination , and basic faults, e.g. bearing noise and damage, balance, looseness and misalignment
Compressors	Recognize examples of centrifugal and screw compressors
Rolling mills, paper machines, other process equipment	Recognize examples of centrifugal and screw compressors of these machines
Machine tools	Recognize examples of these machines
Structures, piping	Recognize the term: resonance

FORMULA SHEET FOR CAT I PREP I PART 6

T * RPM = GMF =gmf = t * rpm	Gear box
$BPF \approx VPF \approx \#\, blade(\, or\ vanes) * rpm$ $BPF = Blade\ pass\ frequency\ (order\ or\ cpm)$	Pump, Compressor and Fan
$Synch\ speed\ (Hz) = 2\ LF\ (Hz) \div poles$ $Actual\ Speed\ (rpm) = Sych\ speed\ (rpm) - Slip\ (rpm)$	Motor
$VPF = Vane\ Pass\ Frequency\ (order\ or\ cpm)$ $Order : n = \frac{f}{Speed}$ $f = vibartion\ frequency\ (cpm)$ $Speed = Shaft\ Speed\ (rpm)$	General

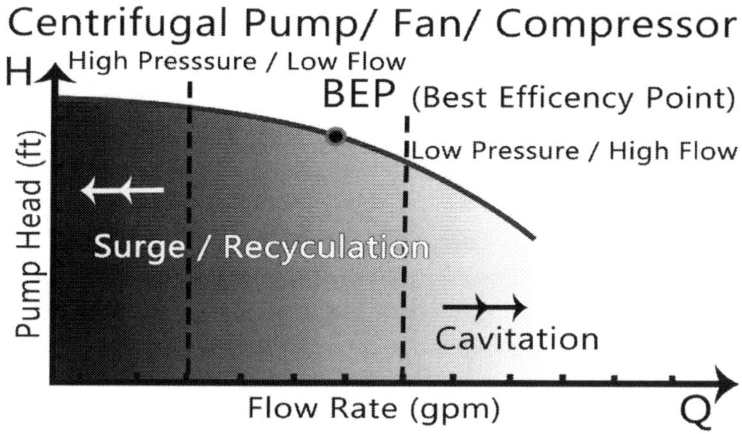
Centrifugal Pump/ Fan/ Compressor
H
High Presssure / Low Flow
BEP (Best Efficency Point)
Low Pressure / High Flow
Pump Head (ft)
Surge / Recyculation
Cavitation
Flow Rate (gpm)
Q

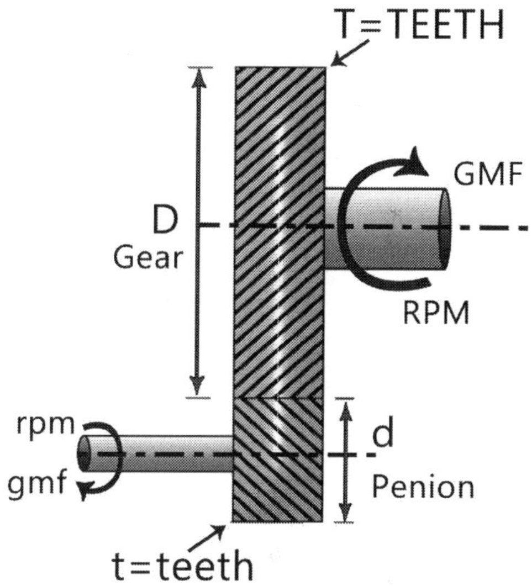
T=TEETH
GMF
D
Gear
RPM
rpm
gmf
d
Penion
t=teeth

Notes

More than

140

Questions

PART

6

2020

QUESTION BANK

Rotating Equipment

Ali M. Al-Shurafa

Rotating Equipment
Question 001

What kind of equipment or part of equipment is illustrated below?

	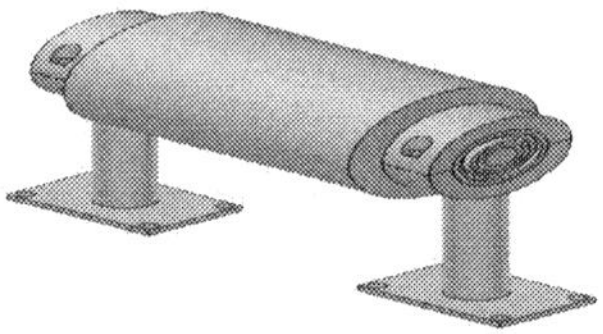	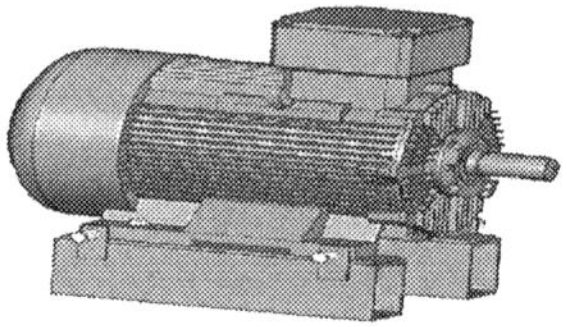
A.	Rolling Mill	AC Induction Motor
B.	Fan's Impeller	Paper Machine
C.	Gearbox	Diesel Engine
D.	Online Vibration Monitoring System	Electric Generator

Rotating Equipment
Question 002

What kind of equipment or part of equipment is illustrated below?

	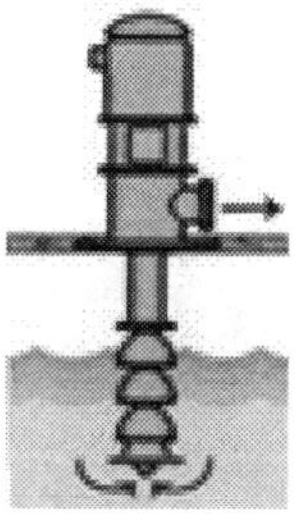	
A.	Vertical Pump	Motor Driven Compressor
B.	Screw Compressor	Belt Conveyer
C.	Belt Conveyer	Pump Filter
D.	Electric Generator	Power Supply Junction Box

Rotating Equipment
Question 003

What kind of equipment or part of equipment is illustrated below?

	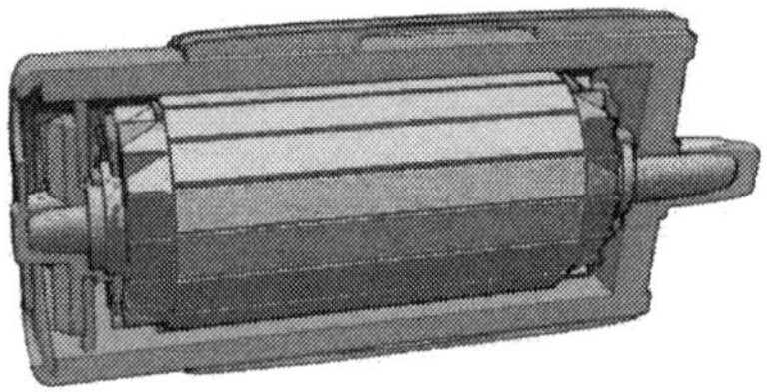	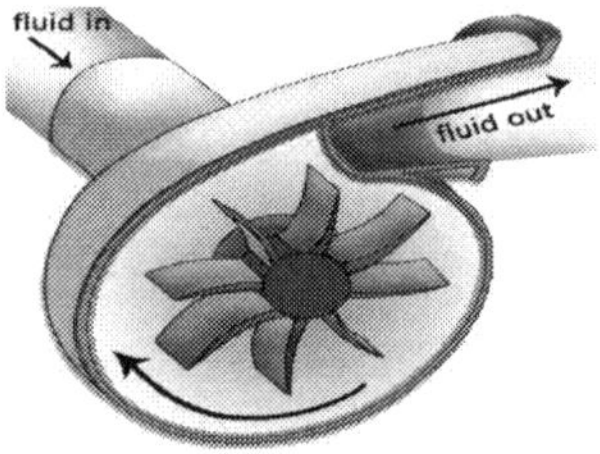
A.	Steam Turbine	Key Phasor
B.	Paper Machine	Multistage Compressor
C.	Screw Compressor	Hanning Window
D.	Electric Motor (cutaway)	Centrifugal Pump

Rotating Equipment
Question 004

What kind of equipment or part of equipment is illustrated below?

	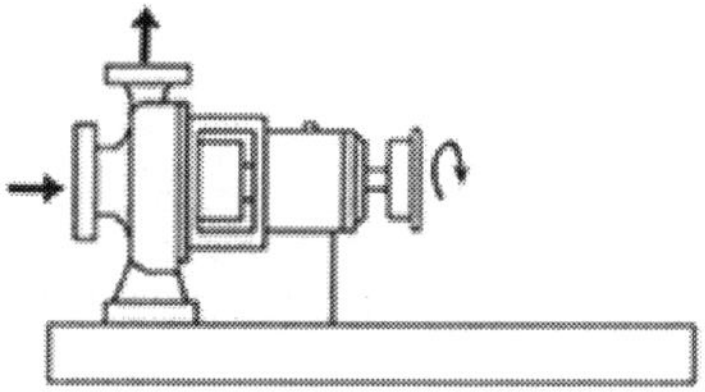	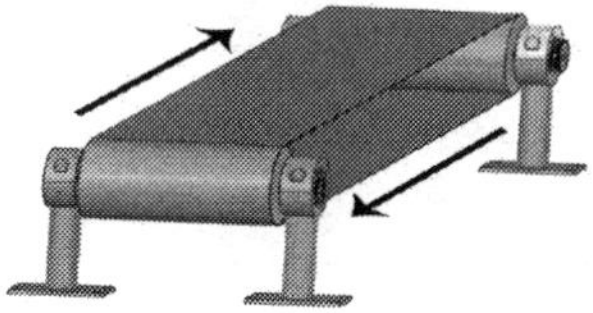
A.	Overhang Pump	Belt Conveyer
B.	Reciprocating Pump	Screw Compressor
C.	Rolling Mill	Anti-aliasing Filter
D.	Electric Generator	High Speed Compressor

Rotating Equipment

Question 005

What kind of equipment or part of equipment is illustrated below?

	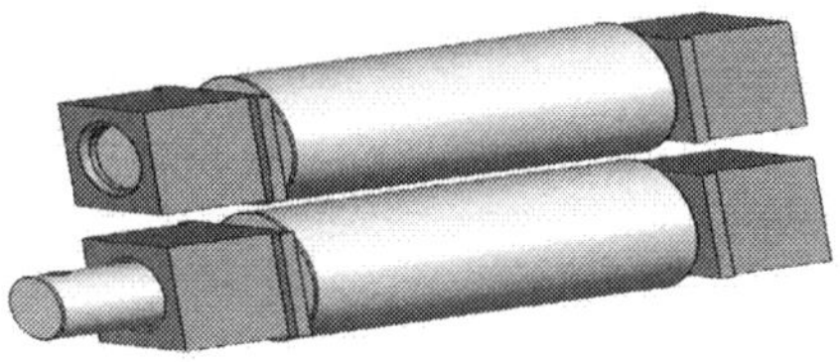	
A.	Rolling Mill	Rolling Element Bearing (exploded)
B.	Between Bearing Pump	Mechanical Seal
C.	Screw Compressor	Vertical Pump
D.	Frequency Spectrum Generator	Journal Bearing

Rotating Equipment
Question 006

What kind of equipment or part of equipment is illustrated below?

		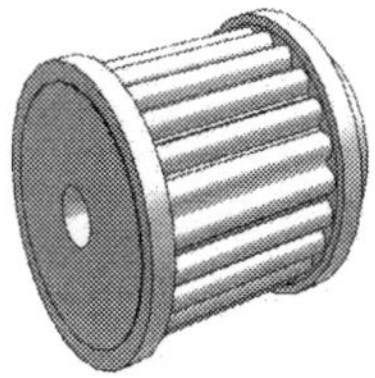
A.	Laser Alignment System	Reciprocating Compressor
B.	FFT Analyzer	Paper Machine
C.	Vertical Pump	Pulley
D.	Data Collector	Vibration Analyzer

Rotating Equipment
Question 007

What kind of equipment or part of equipment is illustrated below?

A.	Belt Conveyer	High Pass Filter
B.	Press Machine	AC to DC Convertor
C.	Vertical Motor	Overhung Pump
D.	Steam Turbine	DC Remover

Rotating Equipment
Question 008

What kind of equipment or part of equipment is illustrated below?

	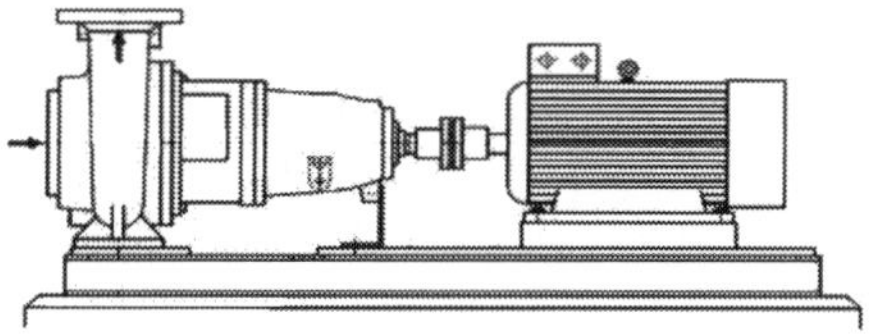	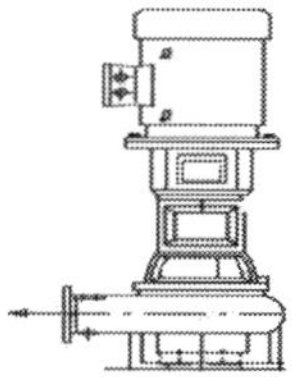
A.	Skid of Motor Driving a Pump	Vertical Motor Driving a Pump
B.	Anti-aliasing Filter	Gearbox
C.	Journal Bearing	Forced Draft Fan
D.	Antifriction Bearing	Flexible Coupling

Rotating Equipment
Question 009

What equipment or part of equipment is illustrated below?

	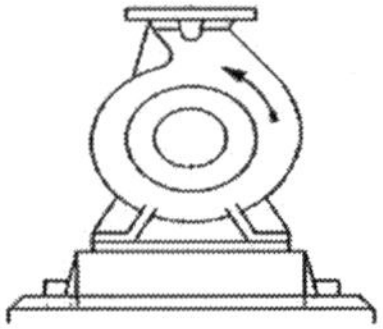	
A.	Rolling Mill	Belt Conveyer
B.	Grinder	Gas Turbine
C.	Centrifugal Pump (side view)	Coupling
D.	Band Pass Filter	Base Plate

Rotating Equipment
Question 010

What kind of equipment or part of equipment is illustrated below?

	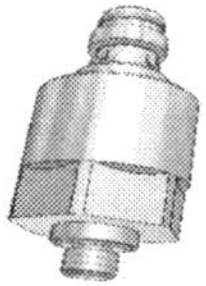	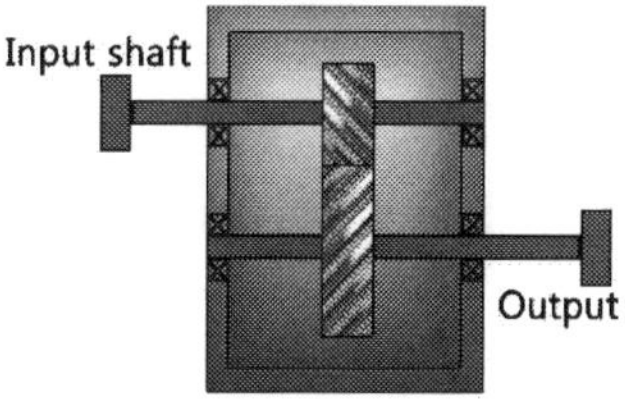
A.	Low Pass Filter	Steam Turbine
B.	Casing Vibration Sensor	Single Stage Gearbox
C.	Paper Machine	Analog to Digital Convertor
D.	Electric Generator	Balancing Machine

Rotating Equipment
Question 011

Classic diesel engines can be described as reciprocating machines. True or False?

A. True, but only if they are connected to a reciprocating compressor or other positive displacement machines.

B. True, regardless of the driven equipment.

C. False, because the output shaft rotates.

D. False, because all drivers belong to the rotating equipment category.

Note : ....

....

....

Rotating Equipment
Question 012

Which group of illustrations corresponds to the type of pumps on the header?

	A.	B.	C.	D.
Overhung		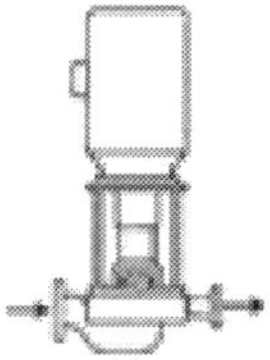	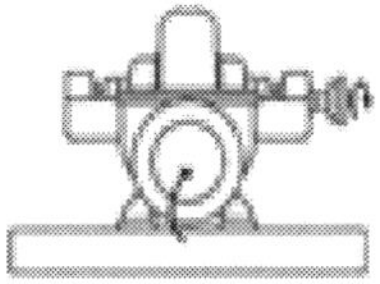	
Suspended	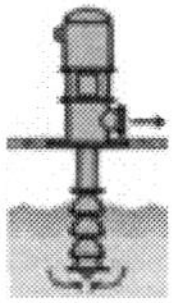	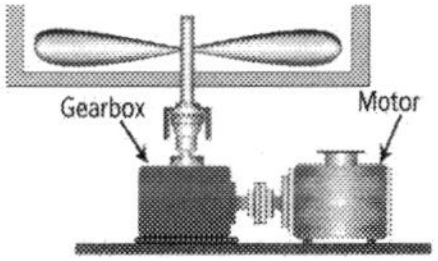	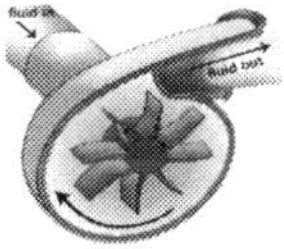	

Note : ....

....

....

Rotating Equipment
Question 013

Which group of illustrations corresponds to the type of pumps on the header?

	A.	B.	C.	D.
Between Bearing	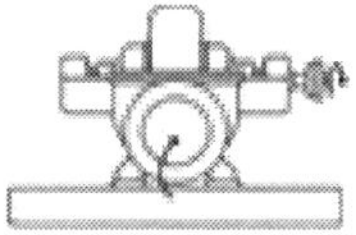	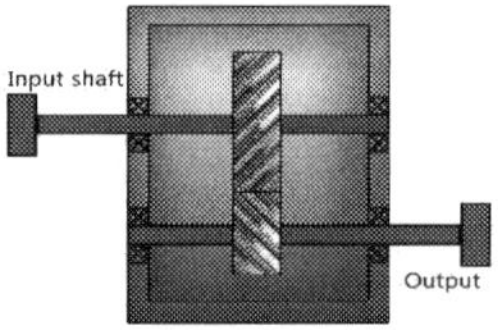	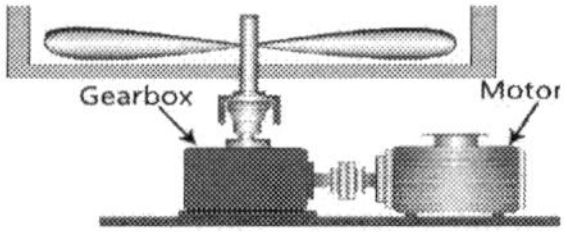	
Inline Pump	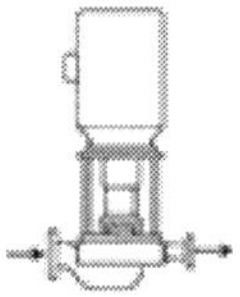		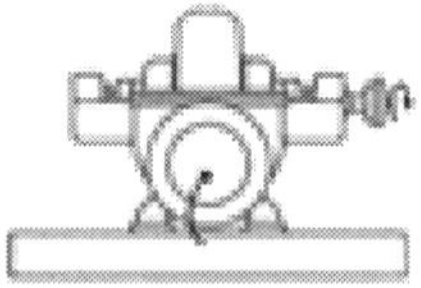	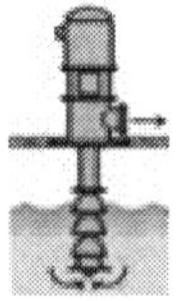

Rotating Equipment
Question 014

Which of the following are called "machine tools"?

A. Wheel grinders

B. Air compressors

C. Seawater pumps

D. Diesel engines

Note : ....

....

....

Rotating Equipment
Question 015

In the industry, which of the following units are used for liquid flow rate?

A.	litter/second	gallon/min	m^3/hr
B.	bar	psi	kPa
C.	mil	ips	g
D.	^oF	^oC	K

Note : ....

....

....

Rotating Equipment
Question 016

In the industry, which of the following units are used for pressure?

A.	litter/second	gallon/min	m^3/hr
B.	bar	psi	kPa
C.	mil	ips	gravity
D.	°F	°C	K

Note : ....

....

....

Rotating Equipment
Question 017

Generally speaking, what is a typical unit for each of the following parameters? These parameters are very important to evaluate the energy consumed by an electric equipment.

	Power	Voltage	Current
A.	GPM or l/min	Volt	dB
B.	Volt or Voltage	ips or mm/sec	gpm or l/min
C.	Hp or Watt	Amp	Volt
D.	Hp or Watt	Volt	Amp

Note : ....

Rotating Equipment
Question 018

For many rotating equipment, the consumed power depends on the rotating equipment's speed and torque. Generally speaking, what is a typical unit for each of the following parameters?

	Power	Speed	Torque
A.	Hp or Watt	rpm	lb•ft or N•m
B.	Hp or Watt	ft/min or m/sec	in/sec or mm/sec
C.	lb•ft or N•m	lb/sec or N/sec	Hp or Watt
D.	Amp	rpm	lb or N

Note : ....

....

....

Rotating Equipment
Question 019

What are these parts? What is their function? Hint: these parts are very thin and vary in thickness.

	Part	Function
A.	Rubber couplings	Power transmission
B.	Strain gauges	Measurement of stress and strain
C.	Coupling keys	Power transmission
D.	Alignment shims	Adjusting the height of equipment

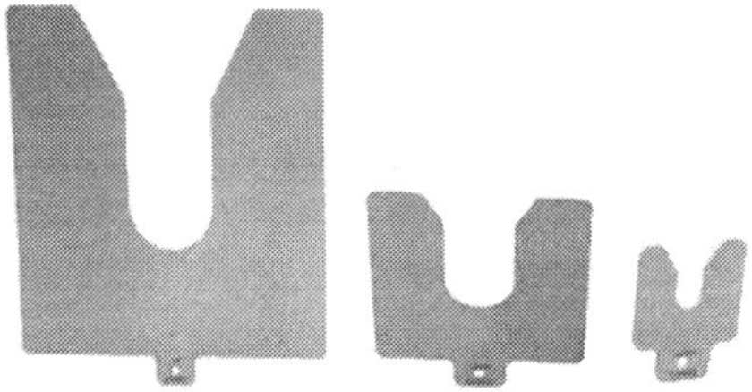

Note : ....

....

....

Rotating Equipment
Question 020

The next illustration shows a part of an industrial pump. What is it? What is its function?

	Part	Function
A.	Nozzle/ Flange	To connect the pump to its inlet and outlet pipes
B.	Shaft	To rotate the bearing
C.	Sleeve	To luck parts on the rotor
D.	Bearing housing	To secure the position of the bearing

Note : ....

Rotating Equipment
Question 021

What is the most common kind of bearings used for rotating equipment in the following industries: printing, paper and rolling mills?

A. Tilting pad bearing

B. Air lubricated bearing

C. Bevel bearing

D. Rolling element bearing

Note : ....

....

....

Rotating Equipment
Question 022

What is a very common kind of bearings usually used for large machines in the following industries: Power Generation, Chemicals, Petrochemicals and Oil & Gas?

A. Journal bearing

B. Grease lubricated bearing

C. Bevel bearing

D. Rolling element bearing

Note : ....

....

....

Rotating Equipment
Question 023

What are the two common types of bearings used for holding shafts in rotating equiment?

A. Journal Bearing and Sleave Bearing

B. Fluid Film Bearing and Rolling Elment Bearing

C. Spure Bearing and Hilical Bearing

D. Rolling Element Bearing and Anti-friction Bearing

Note :

Rotating Equipment
Question 024

What is a function of a radial bearing in rotating equipment? Refer to the illustrations.

A. It allows the rotational motion of a shaft and minimizes its translational motions.

B. It holds the parts of the rotor and rotates them.

C. It converts the thermal/hydraulic energy to rotational energy and vice versa.

D. All of the above.

Rotating Equipment
Question 025

What is a main function of grease for rotating equipment (like a pump) whose shaft is mounted on ball bearings?

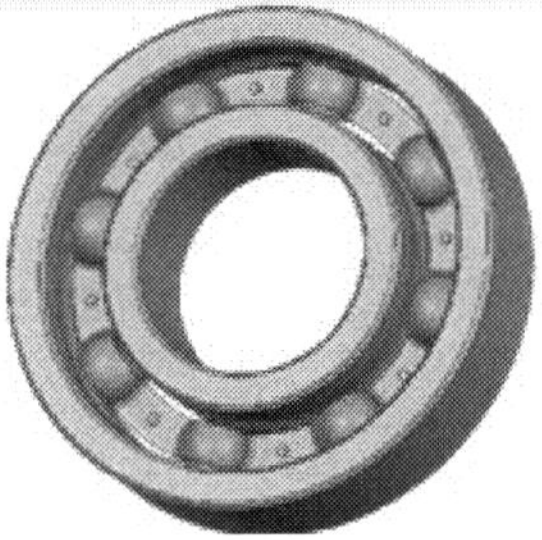

A. Grease delays corrosion on the equipment's case.

B. Grease converts fluid to energy.

C. Grease minimizes and controls the heat of the fluid inside the equipment.

D. Grease reduces friction in the bearings.

Note : ....

....

....

Rotating Equipment
Question 026

Refer to the next simplified illustration for a ball bearing. What are the parts indicated by Arrows 1 and 2?

	1	2
A.	Cage	Ball
B.	Snap ring	Cage
C.	Pinion	Inner race
D.	Inner race	Outer race

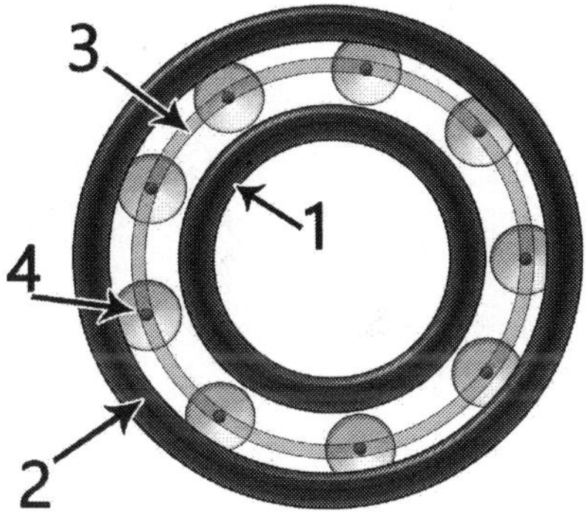

Note : ....

Rotating Equipment
Question 027

Refer to the next simplified illustration for a ball bearing. What are the parts indicated by Arrows 3, 4 and 5?

	3	4	5
A.	Damper	Impeller	Diaphragm
B.	Pinion	Vane	Cage
C.	Cage	Ball	Bearing Housing
D.	Ball	Blade	Volute

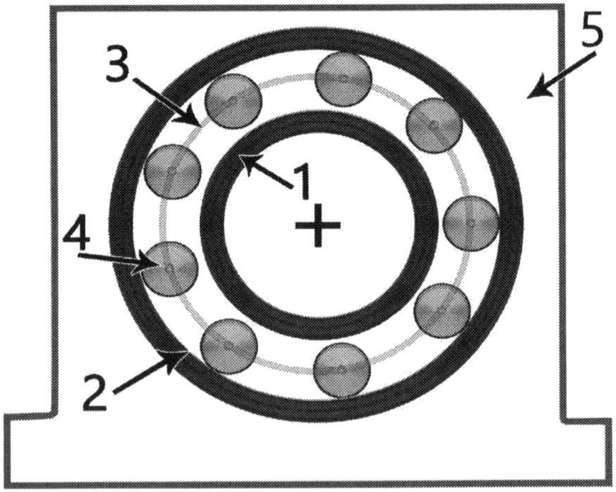

Note : ....

....

....

Rotating Equipment
Question 028

What is another term that can be used to refer to a "sleeve bearing"?

A. Journal bearing

B. Fluid film bearing

C. Antifriction bearing

D. A and B

Note : ....

....

....

Rotating Equipment
Question 029

Refer to the simplified sketch for a tilting pad bearing that is holding a rotor. What does each of the following numbers indicate?

	1	2	3
A.	Shaft	Shaft Centerline	Oil Film
B.	Ball	Pad Gap Voltage	Cage
C.	Pad	Thrust Bearing	Air Gap
D.	Vibration Probe	Probe Centerline	Winding

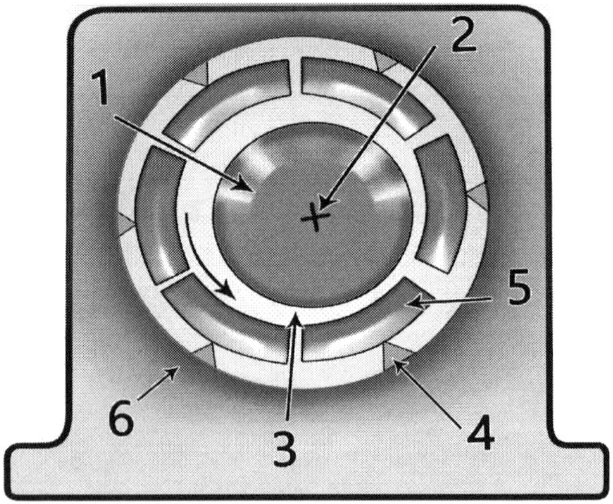

Rotating Equipment
Question 030

Refer to the simplified sketch for a radial tilting pad bearing that is holding a rotor. What does each of the following numbers indicate?

	4	5
A.	Compressor Blade	Roller
B.	Probe Housing	Oil Film
C.	Oil Film	Bearing Housing
D.	Bearing Housing	Bearing Pad

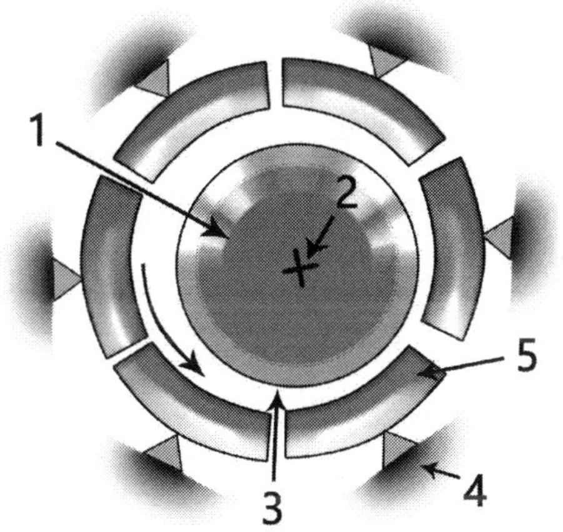

Note : ....

Rotating Equipment
Question 031

What are the main parts of a simple ball bearing? Refer to the illustration.

A. Upper shell, lower shell, balls and liner

B. Balls, rollers, needles and spheres

C. Driver hub, driven hub, shims and spacer

D. Inner race, outer race, cage and rolling elements

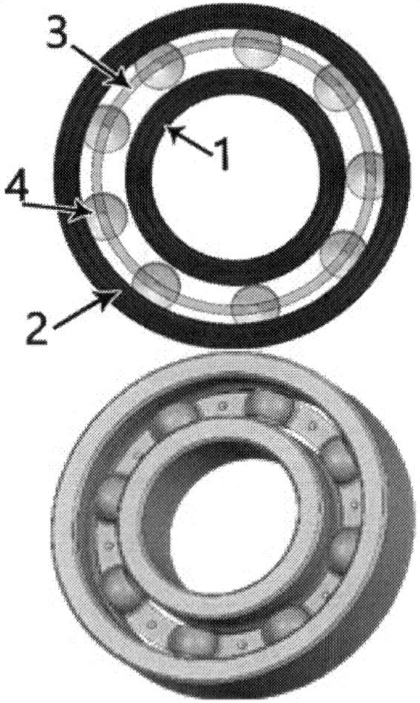

Rotating Equipment
Question 032

What is another term that can be used to refer to a "ball bearing"?

A. Antifriction bearing

B. Roller bearing

C. Rolling element bearing

D. A and C

Note : ....

....

....

Rotating Equipment
Question 033

For an axial vibration measurement on machines, it is recommended to put the sensor on the load zone. What is true about "load zone"?

A. It is the lowest point in the horizontal, vertical and axial directions on the non-drive end bearing.

B. It is an axial area of a bearing (or a bearing housing) at which the maximum axial load is expected to be on the bearing.

C. It is a point on the machine where the highest vibration is measured.

D. It is an imaginary location on a bearing housing that generates FFT spikes.

Rotating Equipment
Question 034

Refer to the sketch for a horizontal pump driven by a motor (not shown). Which location identifies the inboard bearing of this between-bearing pump?.

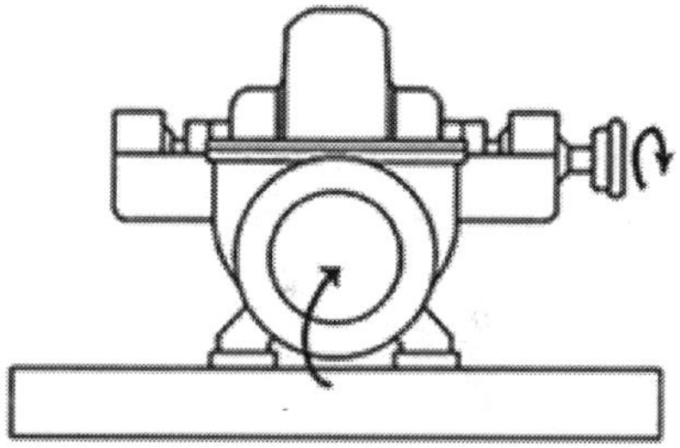

A. Bearing next to the coupling.

B. Bearing next to the suction pipe.

C. Bearing next to the discharge pipe.

D. Bearing above the motor's cooling fan.

Note : ....

Rotating Equipment
Question 035

What drivers are used to operate pumps and compressors in the industry?

A.	Motors	Steam turbines	Gas turbines
B.	Generators	Gearboxes	Circuit breakers
C.	Coupling hubs	Coupling guards	Coupling spacers
D.	Proximitors	Communication gateways	Card isolators

Note : ....

....

....

Rotating Equipment
Question 036

Both motors and generators are electric rotating equipment. What is a major difference between motors and generators?

	Motors ________________ .	Generators ________________ .
A.	consume electricity.	produce electricity.
B.	consume rotational motive power.	consume electricity.
C.	produce rotational motion.	consume rotational motive energy.
D.	A and C	

Note : ....

....

....

Rotating Equipment
Question 037

What are three main parts of a typical electric motor?

A.	Rotor	Stator	Bearings
B.	Plunger	Inlet valve	Outlet valve
C.	Pressure gauge	Temperature gauge	Flow meter
D.	Shaft	Impeller	Seal

Note : ....

Rotating Equipment
Question 038

What part of electric motor is magnified in the next illustration? What is its function?

	Part	Function
A.	Gearbox	To increase the speed
B.	Packing/seal	To control leakage from bearing
C.	Coupling guard	To prevent staff from touching rotating parts
D.	Power junction box	To enable connecting and disconnecting electric cables

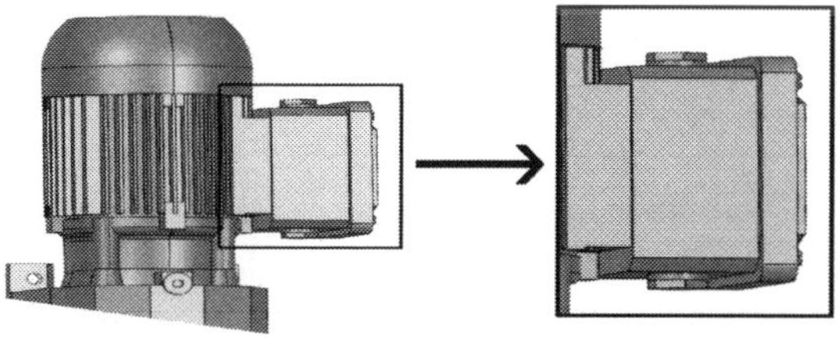

Rotating Equipment

Question 039

In the modern industry, what are the most commonly used power frequencies for electric motors?

A. 50 Hz and 60 Hz

B. 50 cpm and 60 cpm

C. 1500 cpm and 3600 cpm

D. 3 phase and 360 deg (AC and DC)

Note : ....

....

....

Rotating Equipment
Question 040

For a vibration analyst, what is a key difference between an induction motor and a synchronous motor? An induction motor is sometimes called "asynchronous" motor.

	An induction motor ____________	A synchronous motor ________
A.	has a slip.	has no slip.
B.	has no excitation system.	has an excitation system.
C.	is available in a vertical and horizontal layouts.	is only available in a vertical layout.
D.	A and B	

Note : ....

....

....

Rotating Equipment
Question 041

For industrial motors, standard name plates typically contain key technical data about the motor such as ___________ .

A. speed, suction and discharge pressures / heads

B. input power (volt, current, phases etc.)

C. output power (volt, current, phases etc.)

D. B and C

Note : ....

....

....

Rotating Equipment
Question 042

In the field of condition monitoring, what are "poles" when electric motors and generators are discussed?

A. The two extreme locations on Earth (South and North).

B. North and south sides of the magnetic field developed during the operation of these equipment.

C. The number of bearings required to mount the rotor (of motor or generator). They come in pairs and reduce vibration.

D. The grounding cables used to earth (remove electricity) from the case of these machines.

Note : ....

....

....

Rotating Equipment
Question 043

Study the following simplified sketches for electric motors. Which sketch represents a four (4) pole motor?

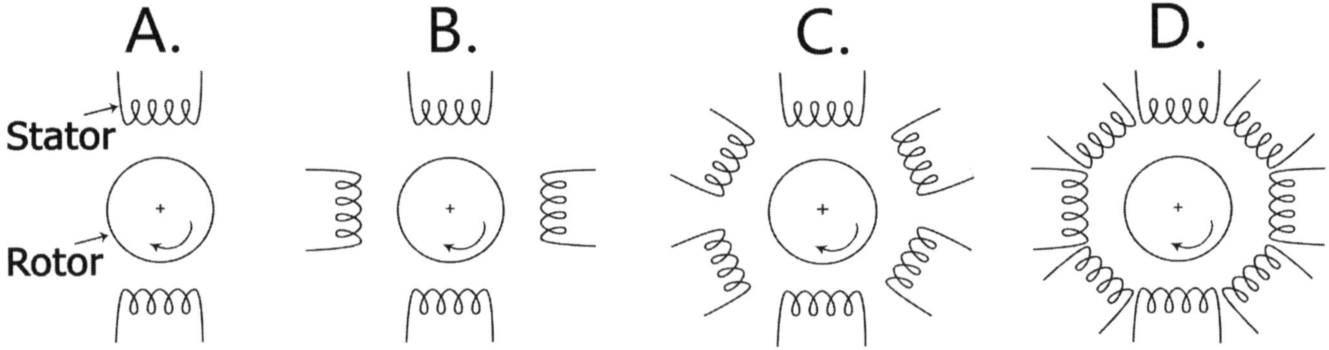

Note : ....

....

....

Rotating Equipment
Question 044

For electric motors and generators, what is true about "air gap"? Refer to the cutaway of a typical motor.

A. It is the difference between the voltage readings of the 3 phases due to poor cable connections.

B. It is the amount of air needed to purge the motor / generator after a shutdown.

C. It is an excessive bearing clearance in journal bearings.

D. It is a radial distance (clearance) between the stator and rotor.

Note : ....

Rotating Equipment
Question 045

What is true about "uneven air gap" for an electric motor?

A. There are different voltage readings among the 3 phases. A typical limit is 10%.

B. It is an excessive axial tension in the coupling that results in uneven deformations and gaps.

C. It is a varying (not uniform) clearance between the stator and rotor. If the variation is large, vibration can develop.

D. It is an excessive clearance in motor's bearings specially with lemon type bearings. If it is large, vibration can develop.

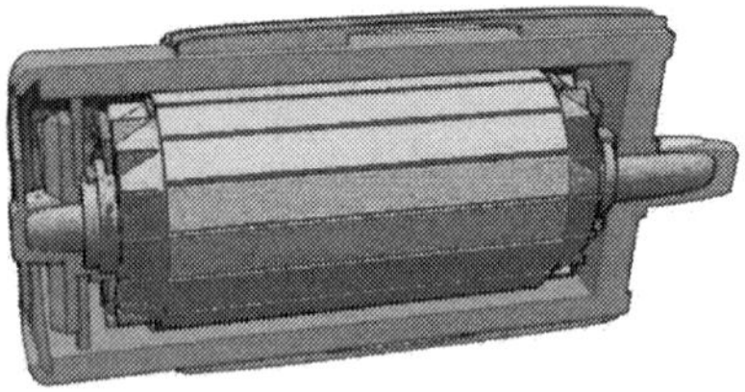

Rotating Equipment
Question 046

In modern industrial plants, many medium-to-large industrial electric motors are 3 phase machines. True or False?

A. True, the power is supplied through 3 poles.

B. True, the power is supplied through 3 electric cables (one for each phase).

C. False, phases must be in pairs. Phases cannot be odd numbers.

D. False, phase value changes as motor crosses its critical speed. Phase is not a fixed number.

Note : ....

....

....

Rotating Equipment
Question 047

Study the nameplate of a three phase induction motor. What is the vibration frequency of 1X and 10X running speed?

	1X	10X
A.	30 Hz	1200 Hz
B.	29.4 cpm	300 Hz
C.	60 Hz	29. 15 Hz
D.	1765 cpm	294.17 Hz

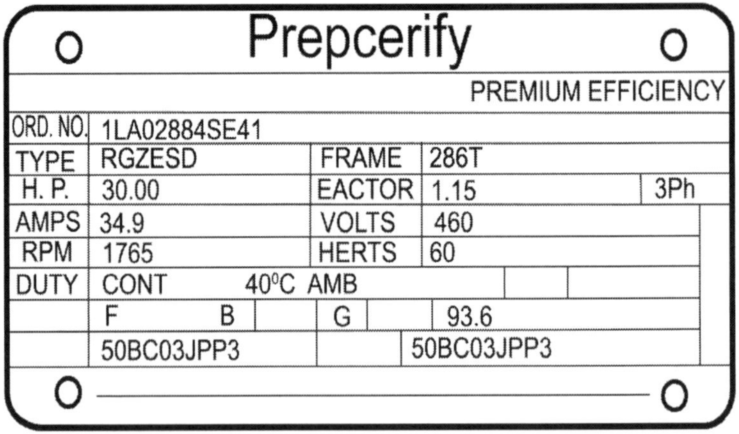

Note : ....

Rotating Equipment
Question 048

Study the next motor nameplate. What is the vibration frequency of 1X running speed? What is the electric power consumed?

	1X	Power
A.	50 cpm	5.21 A
B.	50 Hz	1 HP
C.	1450 cpm	0.75 kW
D.	3000 cpm	230 Volt

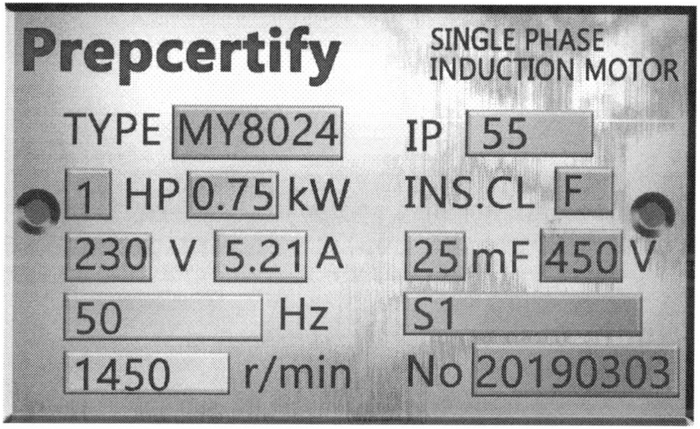

Note : ....

....

....

Rotating Equipment

Question 049

Study the nameplate of a three phase induction motor. What is the vibration frequency of a rotor unbalance? What is the frequency of the electric power supply?

	Unbalance	Power Frequency
A.	30 Hz	120 Hz
B.	29.15 cpm	30.00 Hz
C.	60 Hz	34.9 Hz
D.	1765 cpm	3600 cpm

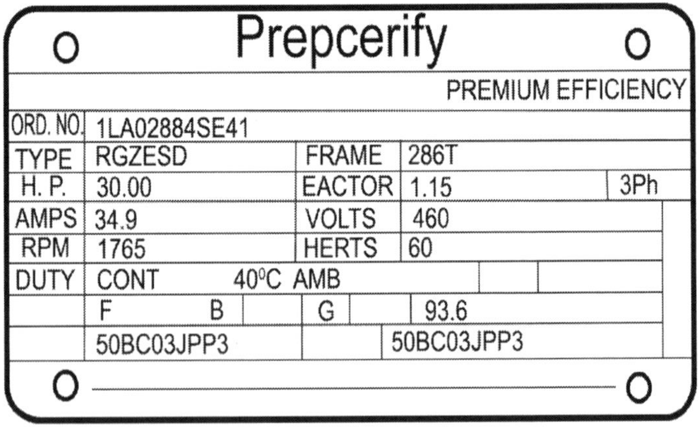

Note :

Rotating Equipment
Question 050

Study the next motor nameplate. What is the vibration frequency of 2X? What is the frequency of the electric power supply?

	2X	Electric Frequency
A.	25 Hz	450 V
B.	100 Hz	24.17 rpm
C.	1450 cpm	5.21 A
D.	2900 cpm	3000 cpm

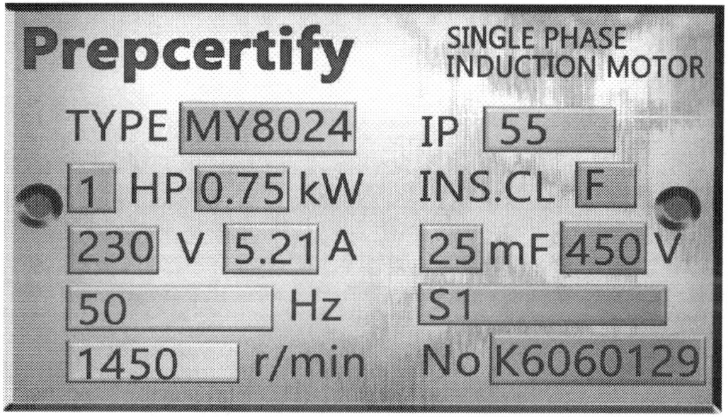

Note : ....

....

....

Rotating Equipment
Question 051

Study the next motor nameplate. Assume that the motor runs at 50% of the full load. At this load, will the voltage (supplied to the motor) drop to 50 %? Will the line frequency drop to 50%?

	Voltage	Frequency
A.	Yes	Yes
B.	Yes	No
C.	No	Yes
D.	No	No

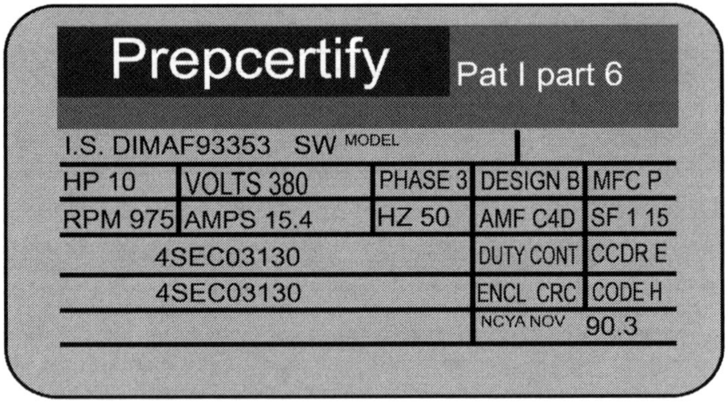

Note : ....

....

....

Rotating Equipment
Question 052

Study the next motor nameplate. Assume that the motor runs at 50% of the full load. At this load, will the efficiency drop to 50 %? Will the speed drop to 50%?

	Efficiency	Speed
A.	Yes	Yes
B.	Yes	No
C.	No	Yes
D.	No	No

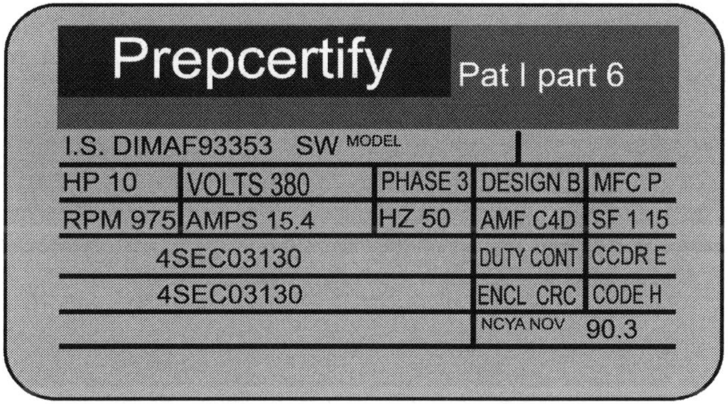

Note :

Rotating Equipment
Question 053

Study the next motor nameplate. Assume that the motor runs at 50% of its full load. What is the absolute frequency of 1X? What would be the consumed current?

	1X Frequency	Current
A.	Between 975 and 1000 cpm	Less than 15.4 amperes
B.	Between 1000 and 975 cpm	More than 15.4 amperes
C.	3000 cpm	50 Hz
D.	More than 50 Hz	More than 10 HP

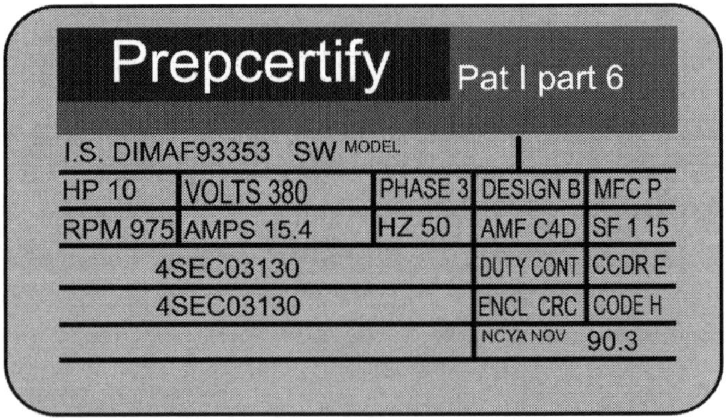

Rotating Equipment
Question 054

Consider medium to large (10-1000 HP) electric AC 3 phase motors in the modern industry. In what range is a typical efficiency of such a motor? What is a typical range of the full load amperage? Select the best option from the given answers.

	Efficiency	Current
A.	10 - 30 %	0.04 - 4 mAmp
B.	30 - 60 %	0.04 - 4 Amp
C.	60 - 80 %	40 - 400 mAmp
D.	80 - 98 %	10 - 100 Amp

Note : ....

....

....

Rotating Equipment
Question 055

What is true about an electric power generator and a gas turbine?

	An electric power generator _____.	A gas turbine _________ .
A.	produces rotational power to be used by other equipment	increases pressure and transfers gas
B.	produces electricity	produces rotational power
C.	increases the torque of the shaft to be used by other equipment	converts electric power from AC to DC
D.	increases pressure and transfers liquid	changes shaft speed, direction and torque

Note : ....

Rotating Equipment
Question 056

What is true about a classic electric motor?

	An electric motor ________ .
A.	changes input speed, shaft centerline direction and torque
B.	increases pressure and transfers liquid
C.	produces rotational power
D.	produces electric power

Note : ....

....

....

Rotating Equipment
Question 057

For industrial induction motors, what is a ''slip''?

A. Motor trip (or shutdown) due to motor over speed (only during startup).

B. The difference between the shaft's actual operating speed and the frequency of the motor's magnetic field.

C. The risk of a motor damage due to its container slipping and getting damaged during the shipment.

D. A design constant related to speed increase in motors that are lightly loaded. Common Example: 6 Hz for 60 Hz motors.

Note : ....

....

....

Rotating Equipment
Question 058

What is the slip for a synchronous motor (10,000 HP) running under 100 % of its rated/design full load? Assume power line is 50 Hz. Hint: Slip is found indirectly by measuring the running speed.

A. Less than 1 rpm (Zero).

B. Around 10 rpm.

C. Greater than 100 rpm.

D. The slip varies a lot based on the number of poles and type of radial bearings (internal resistance).

Note : ....

....

....

Rotating Equipment

Question 059

What is the most reasonable or typical slip value for the following case? A two pole, 60 Hz, horizontal induction motor (200 HP or 150 kW) running at 100% of its design load. Hint: Synchronous speed is 3600 rpm.

A. 0.1 rpm

B. 1 rpm

C. 10 rpm

D. 100 rpm

Note : ....

....

....

Rotating Equipment
Question 060

What is the most reasonable or typical slip value for the following case? A four pole, 50 Hz, vertical induction motor (200 HP or 150 kW) running at 100% of its design load. Hint: Synchronous speed is 1500 rpm.

A. 0.1 rpm

B. 1 rpm

C. 10 rpm

D. 100 rpm

Note : ....

Rotating Equipment
Question 061

What is the formula used to find the synchronous speed of AC electric motors? Hint: Acommon speed is 1800 rpm for a 60 Hz LF.

A. Shaft Speed (Hz)= 2* Line Frequency (Hz) ÷ Number of Poles

B. Shaft Speed (rpm)= 2* Line Frequency (Hz) ÷ Number of Poles

C. Shaft Speed (Hz)= 60* π* Line Frequency (Hz) * Number of Poles

D. Shaft Speed (rpm)= 60* π* Line Frequency (Hz) * Number of Phases

Note : ....

....

....

Rotating Equipment
Question 062

What is the synchronous speed (in rpm) for electric motors whose poles are 2 and 4 if the line frequency is 50 Hz? Hint: use the formula for synchronous speed N = 2LF ÷ poles.

	2 Poles	4 Poles
	Speed (rpm)	
A.	1500	3000
B.	3000	1500
C.	3590	1790
D.	3600	1800

Rotating Equipment
Question 063

What is the synchronous speed (in rpm) for electric motors whose poles are 6 and 8 if the line frequency is 60 Hz? Hint: use the formula for synchronous speed N = 2LF ÷ poles.

	6 Poles	8 Poles
	Speed (rpm)	
A.	1200	900
B.	1195	895
C.	1000	700
D.	900	1200

Rotating Equipment
Question 064

A two-pole induction motor (with a 10 rpm slip) is running on a 60 Hz line frequency. What is the synchronous speed in rpm? What is the actual (operating) speed in rpm? What is the current at the full load?

	Synchronous	Actual	F.L. Current
A.	Unknown	1790	60 Amp
B.	3590	3600	30 Ampere
C.	3600	3590	Unknown
D.	3610	Unknown	600 Amp

Note : ....

Rotating Equipment
Question 065

Consider an AC two-pole, three-phase induction motor running on a 50 Hz line frequency with a 5 rpm slip. What is the frequency of the magnetic field? What is the frequency of 1X shaft speed? Hint: First, find synchronous speed then use slip to find the actual speed.

	Magnetic Field Frequency	1X (cpm)
A.	2 Hz	5
B.	60 Hz	3595
C.	50 Hz	2995
D.	3000 cpm	1495

Rotating Equipment
Question 066

Consider an AC three-phase, four-pole induction motor running on a 60 Hz line frequency with a 10 rpm slip. What is the 2-line frequency for this motor in Hz? What is the motor's frequency of 2X running speed? N = 2LF ÷ poles.

	2 Line Frequency (Hz)	2X Speed (rpm)
A.	60	3590
B.	120	3580
C.	1790	240
D.	7200	1790

Rotating Equipment
Question 067

Consider a standard in your company for electric motors that calls for a maximum slip of 2%. What is the maximum slip in rpm for an induction motor running with 6 poles and on a 480 Volt line? Assume a 60 Hz line frequency. Hint: Find synchronous speed then find the slip based on the standard.

A. 24 rpm

B. 36 rpm

C. 54 rpm

D. 1200 rpm

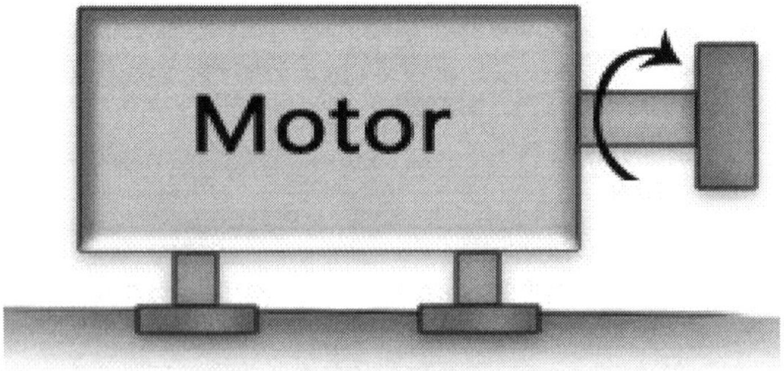

Rotating Equipment
Question 068

Consider a motor standard in your company for electric motors that calls for a maximum slip of 3%. What is the maximum slip in rpm for an induction motor running with 6 poles and on a 480 Volt line? Assume a 50 Hz line frequency. Hint: Find synchronous speed then find the slip based on the standard.

A. 16 rpm

B. 30 rpm

C. 60 rpm

D. 1000 rpm

Rotating Equipment

Question 069

Consider a motor standard in your company that calls for a maximum slip of 4%. Accordingly, what is the minimum acceptable speed for an induction motor you should accept? Assume the motors run with 4 poles and use a 60 Hz electric power. Hint: Synch speed is 30 Hz.

A. 72 rpm

B. 1872 rpm

C. 29.6 Hz

D. 1728 rpm

Note : ....

....

....

Rotating Equipment
Question 070

Consider a motor standard in your company that calls for a maximum slip of 1.5 %. Accordingly, what is the minimum acceptable speed for an induction motor you should accept? Assume the motors run with 2 poles and use a 50 Hz electric power. Hint: N = 2LF ÷ poles.

A. 70 rpm

B. 2955 rpm

C. 3045 rpm

D. 3546 rpm

Note : ....

....

....

Rotating Equipment
Question 071

Consider a motor standard in your company that calls for a maximum slip of 2%. Accordingly, what is the maximum slip in rpm for an induction motor running with 4 poles and on a 4.16 kV power? Hint: Line frequency is not given. Assume an appropriate value. Select the best option from the given answers.

A. 15 rpm

B. 16 rpm

C. 18 rpm

D. 36 rpm

Note : ....

....

....

Rotating Equipment
Question 072

Assume the line frequency is 3000 cpm. Which of the following is a possible magnetic field (synchronous) speed for an AC electric motor? Hint: Use the formula for different poles: N = 2LF ÷ poles.

999 rpm	1111 rpm	1500 rpm	2000 rpm

A. Only 1500 rpm

B. Only 999 and 1111 rpm

C. Only 2000 rpm

D. All of the above

Note : ....

....

....

Rotating Equipment
Question 073

Which of the following is a possible magnetic field (synchronous) speed for a shaft of an AC electric motor? Assume line frequency is 3600 cpm. Hint: Use the formula for different poles: N = 2LF ÷ poles.

15 Hz	60 Hz	60.5 Hz	120 Hz

A. Only 15 and 60 Hz

B. Only 60 and 60.5 Hz

C. Only 15, 60 and 120 Hz

D. None of the above

Note : ....

....

....

Rotating Equipment
Question 074

Which of the following is a possible synchronous vibration frequency for an AC electric motor? Assume line frequency is 60 Hz. Hint: Use the formula. Motor speed in the question is unknown. Synchronous frequency is an exact integer of the speed. Non-synchronous is any speed below the synch speed or non-integer multiple of the speed.

900 rpm	1200 rpm	1500 rpm	1508 rpm	2988 rpm	3000 rpm

A. Only 1500 rpm

B. Only 1500 and 3000 rpm

C. Only 1508 and 2988 rpm

D. Only 900 and 1200 rpm

Rotating Equipment
Question 075

Which of the following is defiantly a NON-SYNCHRONOUS frequency for a vibration signal? Assume a synchronous motor running on a 60 Hz line frequency. Hint: Motor's speed in the question is unknown. Synchronous frequency is an exact integer for speed. Non-synchronous is any frequency below the synch speed or non-integer multiple of the speed.

15 Hz	20 Hz	25 Hz	30 Hz
31 Hz	50 Hz	59.5 Hz	60 Hz

A. Only 15, 30 and 60 Hz

B. Only 25 and 50 Hz

C. Only 20 and 31 Hz

D. Only 25, 31, 50 and 59.5 Hz

Rotating Equipment
Question 076

Which of the following is a possible shaft speed for an AC electric motors (synchronous or induction)? Assume line frequency is 60 Hz. Hint: The poles of the motor are unknown. Use the formula. Assume reasonable slip.

2990 rpm	3200 rpm	3600 rpm	6000 rpm
50 Hz	59.75 Hz	62 Hz	120 Hz

A. Only 3600 rpm and 59.75 Hz

B. Only 3600 and 6000 rpm

C. Only 2990 rpm, 3600 rpm, 50 Hz and 62 Hz

D. Only 2990 rpm, 3590 rpm, 62 Hz and 120 Hz

Rotating Equipment

Question 077

In the current industry (year 2020), do manufactures produce standard induction motors running at 1360 rpm?

A. Possible and common, if line frequency is 50/60 Hz and the stator has 13 poles.

B. Possible if line frequency is 50 Hz (3000 cpm) and motor is 3 phase.

C. Possible if line frequency is 50 Hz (3000 cpm) and motor is 3 poles.

D. Speed at 1360 rpm is technically impossible based on 50 or 60 Hz electricity.

Rotating Equipment
Question 078

What are three main parts of a typical centrifugal compressor?

A.	Rotor	Winding	Bearings
B.	Plunger	Inlet valve	Outlet valve
C.	Pressure gauge	Temperature gauge	Flow meter
D.	Shaft	Impeller	Seal

Note : ....

Rotating Equipment

Question 079

Which part exists in typical centrifugal compressors but does not exist in typical screw compressors?

A. Seal

B. Impeller

C. Bearing

D. Shaft

Note : ....

....

....

Rotating Equipment
Question 080

Consider a classic horizontal medium size centrifugal pump. At what speed does the impeller rotate? Refer to the illustration.

A. At half times the shaft's speed.

B. At the same speed of the shaft.

C. At shaft's speed times the number of vanes in the impeller.

D. It depends on the critical speed.

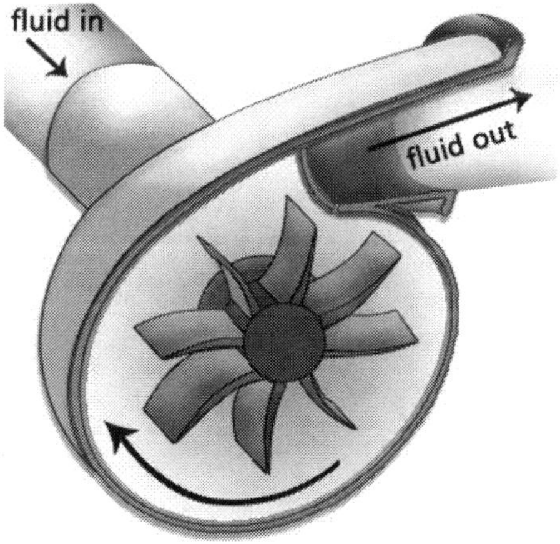

Note : ....

....

....

Rotating Equipment
Question 081

Consider the basic concepts for gas compression systems. What is a key difference between centrifugal and screw compressors?

	Centrifugal Compressors	Screw Compressors
A.	Larger in size	Smaller in size
B.	Variable flow rate over a range of pressure	Constant flow rate over a range of pressure
C.	Handle gases	Handle liquids
D.	Rotate counter clockwise	Rotate clockwise

Note : ....

....

....

Rotating Equipment
Question 082

There are many types of pumps. Which of the following are common pump types in the modern industry?

A. Analog and digital

B. Centrifugal and positive displacement

C. Reciprocating and condensing

D. B and C

Note : ....

....

....

Rotating Equipment
Question 083

There are many types of compressors. Which of the following are common types in the modern industry?

A. AC and DC

B. Centrifugal and positive displacement

C. Synchronous and induction

D. Forced and induced

Note :

Rotating Equipment
Question 084

What is true about a pump and compressor?

	Pump ____________________.	Compressor ___________.
A.	handles liquids.	handles gases.
B.	handles gases.	handles liquids.
C.	develops discharge pressures greater than or equal to 10 times suction pressure.	develops discharge pressures less than or equal to 10 times suction pressure
D.	develops discharge pressures less than or equal to 10 times suction pressure.	develops discharge pressures greater than or equal to 10 times suction pressure.

Note :

Rotating Equipment
Question 085

What is true about the term "incompressible" when we talk about fluids handled by pumps?

A. The density of the service does not change as it is pressurized. Example is lube oil. This term is important to differentiate between gases and liquids.

B. The vibration data is too big and cannot be saved in a normal size file. Example is a startup file for a large steam turbine.

C. The fluid inside the pump cannot generate cavitation. Example is paint.

D. The fluid leaving the pump cannot be pressurized any more. Example is gasoline. This term is very important in the Oil and Gas Industry.

Rotating Equipment
Question 086

What is the main difference between compressors and pumps?

A. Pumps are smaller than compressors (typically 1/10 times).

B. Pumps are designed to handle liquids but compressors are designed to handle gases.

C. Compressors are more critical than pumps. Unlike, pumps, when compressors trip , plant stops production.

D. Compressors are driven by turbines but pumps are driven by motors.

Rotating Equipment
Question 087

What are main internal parts of a typical centrifugal pump?

A.	Rotor Bars	Stator Slots	Winding
B.	Plunger	Inlet Valve	Outlet Valve
C.	Shaft	Impeller	Seal
D.	All of the above		

Note : ....

....

....

Rotating Equipment
Question 088

Which of the following are main parts of a classic centrifugal fan?

A.	Shaft	Impeller	Casing
B.	Rotor	Stator slots	Winding
C.	Pug	Bearing	Check valve
D.	Pressure gauge	Temperature gauge	Flow meter

Note :

Rotating Equipment
Question 089

What are main parts of a typical reciprocating (positive displacement) pump?

A.	Rotor Bars	Stator Slots	Winding
B.	Plunger	Inlet Check Valve	Outlet Check Valve
C.	Shaft	Impeller	Seal
D.	All of the above		

Note : ....

....

....

Rotating Equipment
Question 090

What is the part shown in the next illustration? What is its function? Hint: This part is stationary and installed in many pumps and valves.

	Part	Function
A.	Rubber coupling	Power transmission
B.	Chain	Power transmission
C.	Alignment shims	Adjusting the height of a machine
D.	Packing	Sealing

Note : ....

Rotating Equipment

Question 091

What part of pump is shown in the next illustration? What is its function? Hint: This part rotates.

	Part	Function
A.	Metallic coupling	To connect motor to pump
B.	Rolling element bearing	To rotate the fluid
C.	Impeller	To increase the pressure/ speed of the fluid
D.	Bevel gear	To secure the position of the shaft and reduce thrust

Note : ....

Rotating Equipment
Question 092

What part of pump is magnified in the next illustration? What is its function?

	Part	Function
A.	Coupling	To transmit power
B.	Junction box	To connect and disconnect power cables
C.	Open impeller	To increase the speed of liquid
D.	Equipment nozzle	To connect the pump to its piping

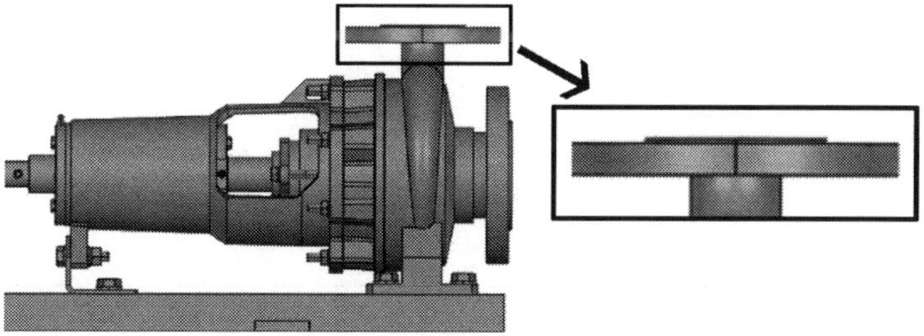

Rotating Equipment
Question 093

What part of pump is shown in the next illustration? What is its function?

	Part	Function
A.	Case	To contain the fluid and direct it to the discharge side
B.	Junction box	To connect and disconnect power cables
C.	Open impeller	To increase the speed of liquid
D.	Equipment nozzle	To connect the pump to its piping

Rotating Equipment
Question 094

What is the function of a nozzle (flange) in a centrifugal pump? Refer to the illustration for an example.

A. It allows rotational motion of a shaft and prevents (or minimizes) its translational motions.

B. It holds the parts of the rotor and rotates them.

C. It converts the thermal/hydraulic energy to rotational energy and vice versa.

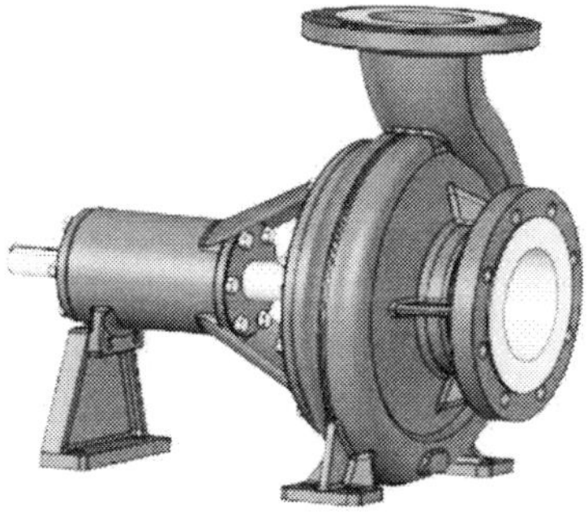

D. None of the above.

Rotating Equipment
Question 095

What is a main function of a shaft key in rotating equipment? Refer to the illustration for an example.

A. It allows the shaft's rotational motion of and prevents its translational motions.

B. It restricts access to the coupling and improves the operational safety.

C. It controls the leakage from the pump to atmosphere.

D. It helps securing the relative locations of rotor's parts and provides a means of power transmission.

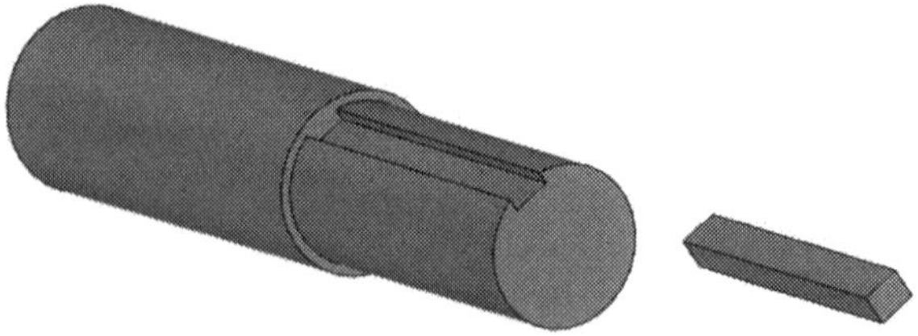

Rotating Equipment
Question 096

What is the primary function of keys and key ways in rotating equipment? Refer to the illustration for an example.

A. They reduce the imbalance of the rotor and as a result they reduce the shaft vibration.

B. They improve the thermal efficiency of the parts mounted on the rotor.

C. They remove any electro-static charges developped in the rotor.

D. None of the above.

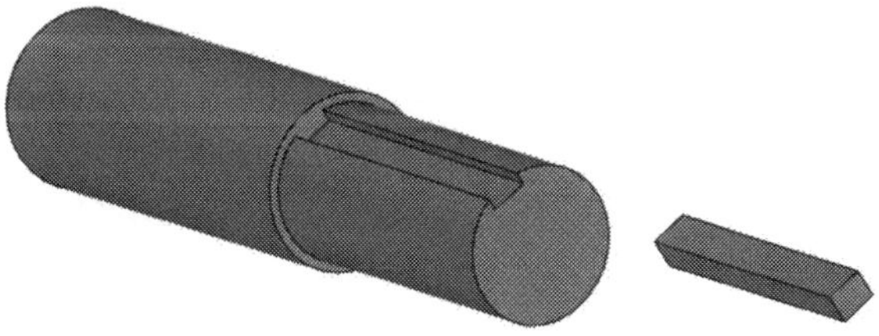

Rotating Equipment
Question 097

What is the function of a shaft-end mechanical seal for a centrifugal pump? Refer to the illustration for an example.

A. It increases efficiency by reducing the internal recirculation.

B. It turns the rotor and holds the parts on it.

C. It contains the vibration inside the pump.

D. It minimizes and controls the leakage of the liquid (service) from the pump to the atmosphere.

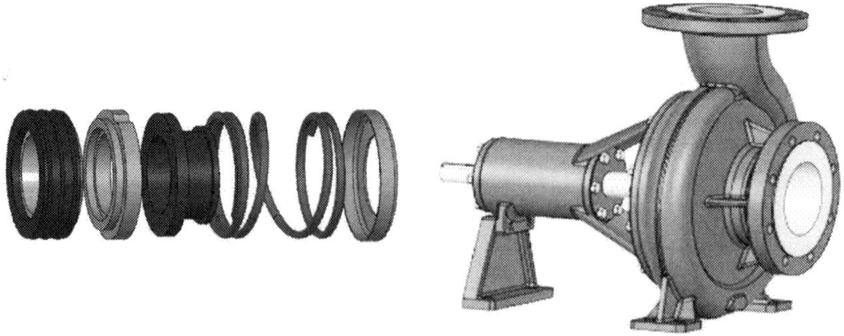

Rotating Equipment
Question 098

What is the function of an impeller for a pump or a fan? Refer to the next images.

A. It counter-acts the thrust force by moving the shaft in the opposite direction.

B. It holds the parts of the rotor and centerlines the rotor.

C. It converts the rotational energy (from the shaft) to fluid energy.

D. It minimizes and controls the internal leakage of the liquid inside the pump or fan.

Rotating Equipment
Question 099

What is a vane or blade in the field of machinery vibration analysis?

A. A signal-processing algorithm, applied specially on pumps when analyzing thier vibrations signals.

B. A hardware used for vibration data collection, applied specially on pumps when analyzing vane pass vibration signals.

C. A part of the impeller in centrifugal pumps or compressors.

D. A mechanical defect of the impeller in centrifugal pump or compressor.

Rotating Equipment
Question 100

Which statement is true about a classic compressor and pump in the current industry?

1- A compressor is used to increase pressure of and transfer gas.
2- A compressor is used to change the shaft speed, direction or torque.
3- A pump is primarily used to produce electric power.
4- Both pump and compressor are primarily used to produce rotational power.

A. Only 1

B. Only 3

C. Only 3 and 4

D. 1, 2, 3 and 4

Rotating Equipment

Question 101

What are typical units for a pump flow rate? Hint: flow rate is sometimes referred to as capacity.

A.	ft /min	rpm
B.	gpm	m^3/hr
C.	psig	kg/cm^2
D.	cpm	Hz

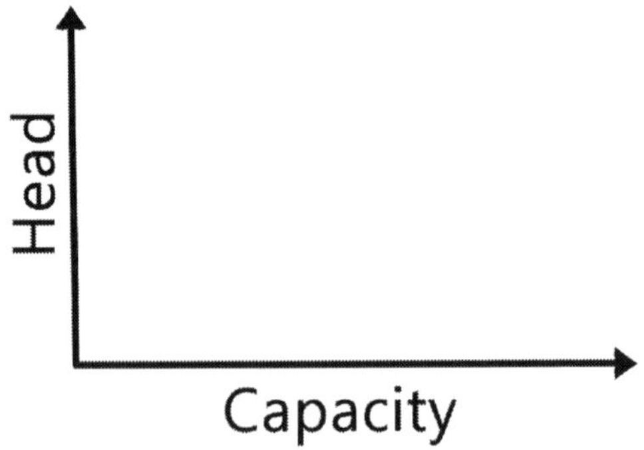

Rotating Equipment
Question 102

What are typical units for a pump head? Hint: head is related to pressure but expressed in different units.

A.	ft	m
B.	gpm	m^3/hr
C.	psig	kg/cm^2
D.	mil	µm

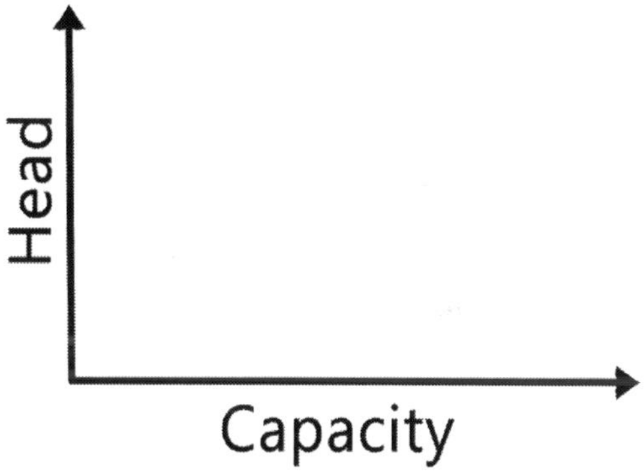

Rotating Equipment
Question 103

What is true about classic steam turbines?

	Turbines______.
A.	increase pressure and transfer steam
B.	produce electric power
C.	produce rotational power
D.	change shaft speed, direction and torque

Note : ....

Rotating Equipment
Question 104

For centrifugal pumps, there is an operating condition referred to as BEP. At this condition, energy losses and hydraulic radial forces reduce to minimum. What does BEP stand for?

A. Best End Play (= shaft OD ÷ 1000)

B. Best Efficiency Point

C. Best Engineered Pumping (It's the point of max flow rate.)

D. Between Ends Point (the opposite of Distance Between Shaft Ends DBSEs)

Note :

Rotating Equipment

Question 105

For centrifugal pumps, the operating performance is evaluated using a graph sometimes called the "characteristic curve". What is the variable in the x-axis, y-axis and their typical units, in this graph? Refer to the next illustration.

	X-axis (unit)	Y-axis (unit)
A.	Flow Rate (gpm or m^3/hr)	Head (ft or m)
B.	Quality (second)	Head (ips or mm/sec)
C.	Frequency (cpm)	Highest Amplitude (ips or mm/sec)
D.	Quantity (gallons or m^3)	Horizontal Vibration (g)

Note : ....

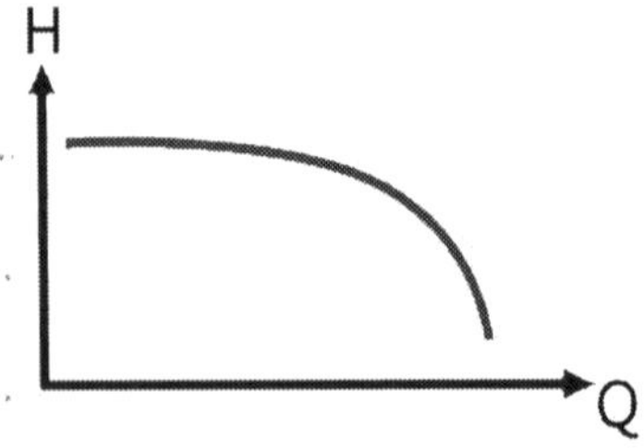

Rotating Equipment
Question 106

For industrial pumps, standard name plates typically contain basic technical data such as _____________ . Select the best option from the given answers.

A. speed, suction and discharge pressures / heads

B. input power (volt, current, phases etc.)

C. output power (volt, current, phases etc.)

D. B and C

Note : ....

....

....

Rotating Equipment
Question 107

Consider a typical performance curve for a pump with a backward curved centrifugal impeller. What is the general relationship between the delivered capacity and pump's head? Hint: Sketch a simple curve to show the relation.

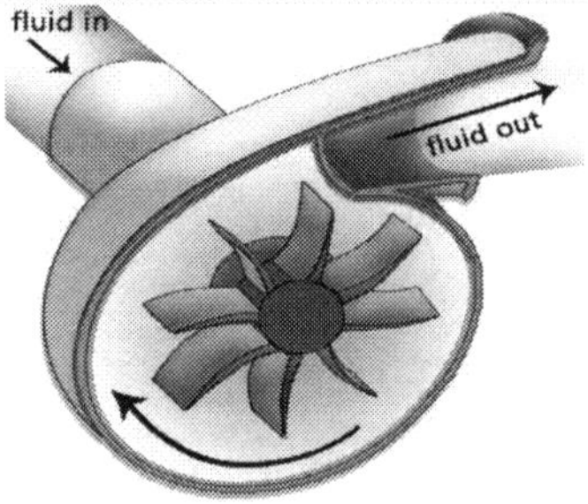

A. As the capacity increases, the head increases.

B. As the capacity increases, the head decreases.

C. As the capacity decreases, the head decreases.

D. As the capacity approaches zero, the head approaches zero.

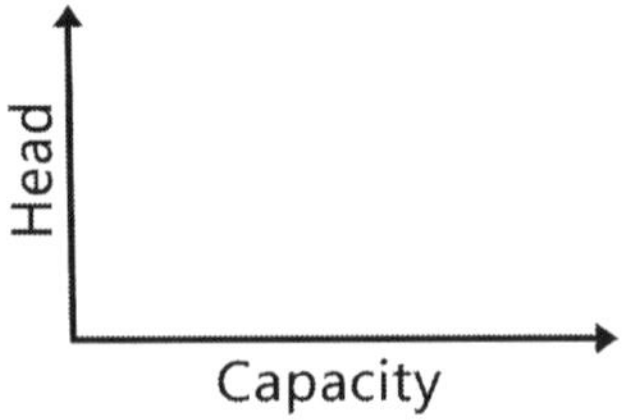

Note : ....

Rotating Equipment
Question 108

The next illustration is for a small pump. Calculate/estimate the blade/ vane passing frequency. The shaft rotates at 995 rpm. The driver is an induction motor (not shown) drawing a 50 Hz electricity.

A. 8 X

B. 16.6 Hz

C. 7960 Hz

D. 59700 cpm

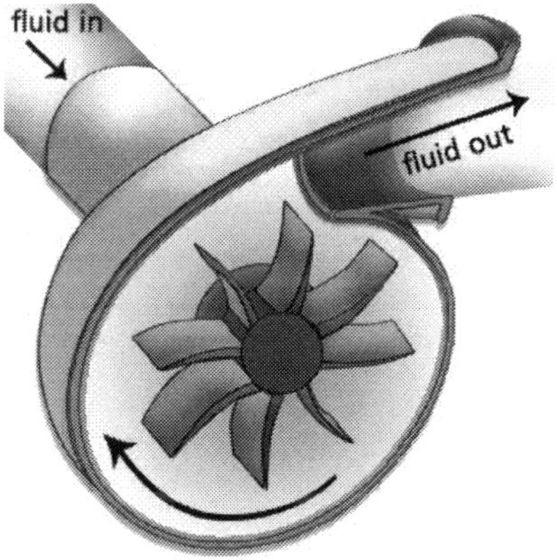

Note : ....

....

....

Rotating Equipment
Question 109

The next illustration is for a small pump. Calculate/estimate the blade/ vane passing frequency. The shaft rotates at 990 rpm. The driver is an induction motor (not shown) drawing a 50 Hz electricity.

A. 124 X

B. 8 cpm

C. 19.8 Hz

D. 132 Hz

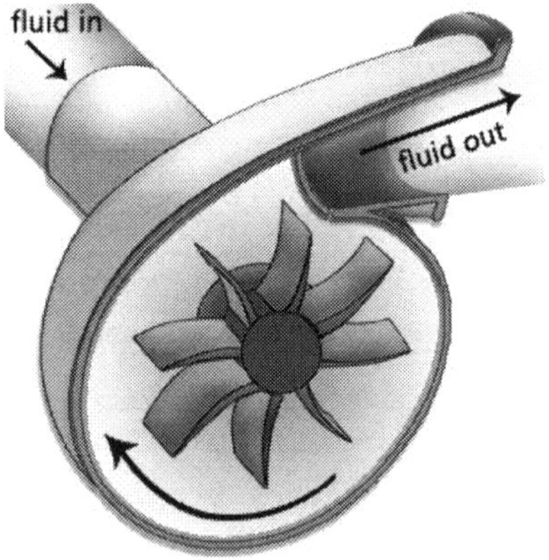

Note : ....

....

....

Rotating Equipment
Question 110

What piece of equipment is shown in the next simplified illustration? Power is 1000 Hp (750 kW).

A. Vertical gearbox (commonly used for paper plants)

B. Centrifugal pump driven by a motor

C. Synchronous motor driving a reciprocating compressor

D. Forced draft fan (commonly used for boilers)

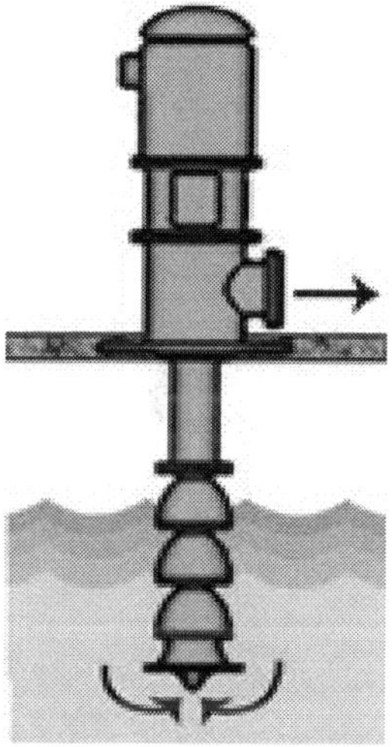

Note : ....

Rotating Equipment
Question 111

What piece of equipment is shown in the next illustration? Power is 50 Hp (37 kW).

A. Horizontal gearbox

B. Centrifugal pump on a skid

C. Synchronous motor driving a fan

D. Forced draft fan with a gearbox

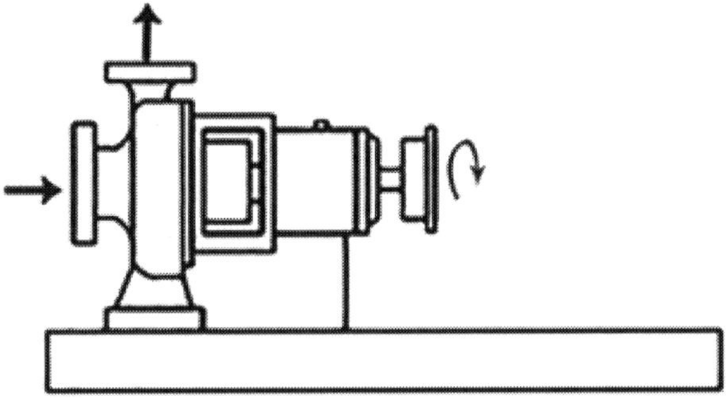

Note : ....

Rotating Equipment
Question 112

What piece of equipment is shown in the next illustration? Power is 500 Hp (373 kW). This type of equipment is common in water, chemicals, power and oil industry.

A. Synchronous motor on a skid

B. Overhung reciprocating pump

C. Between bearing gearbox

D. Between bearing centrifugal pump

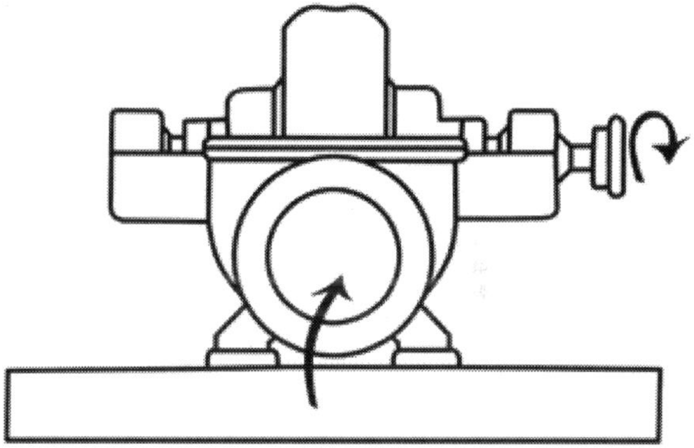

Note :

Rotating Equipment
Question 113

What piece of equipment is shown in the next illustration? Power is 70 Hp (52.2 kW).

A. Inline centrifugal pump driven by a motor

B. Between bearing centrifugal pump

C. Synchronous motor driving a mixer

D. Forced draft fan driven by a gas turbine

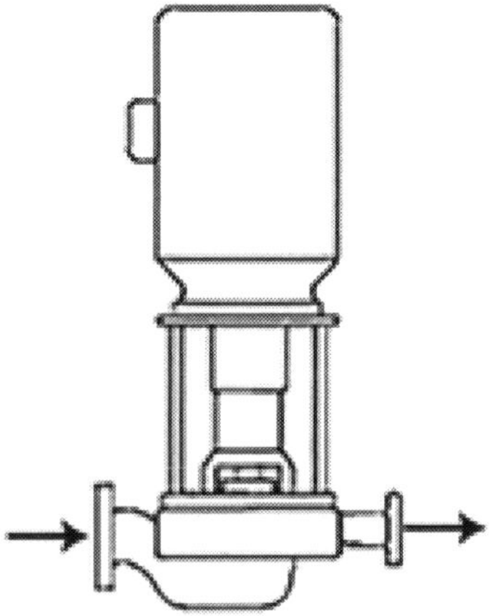

Note : ....

Rotating Equipment
Question 114

What piece of equipment is shown in the photo? Power is 100 Hp (75 kW).

A. A skid of motor-centrifugal pump

B. A skid of gearbox-forced draft fan

C. A skid of generator driven by a motor

D. A skid of diesel engine driven by a motor

Note : ....

Rotating Equipment

Question 115

What piece of equipment is shown in the next photo? Hint: This equipment is common in power generation plants.

A. Atmospheric gearbox

B. Pump driving a fan

C. Centrifugal fan driven by a motor

D. Synchronous turbine

Note : ....

....

....

Rotating Equipment
Question 116

Consider a vertical induction motor running at 2995 rpm. This motor drives a small sump pit pump. Which of the following are examples of synchronous and non-synchronous vibration signals for this pump? Hint: synchronous refers to an exact frequency of shaft speed or its multiples. LF is at 50 Hz.

	Synchronous Vibration Frequency	Non-Synchronous Vibration Frequency
A.	1X and 1500 cpm	2X and 3000 cpm
B.	1X and 2995 cpm	2.01X and 3000 cpm
C.	2995 and 3000 cpm	1X and 2X
D.	50 and 100 Hz	1 LF and 2 LF

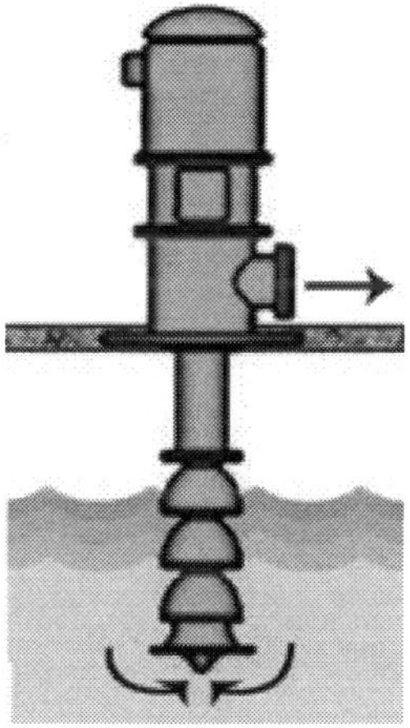

Rotating Equipment
Question 117

For typical centrifugal pumps, cavitation can result in a high vibration and an audible noise. What is a common cause for cavitation?

A. Low suction pressure (below the design limit)

B. High shaft misalignment (> 0.1% of DBSEs)

C. Low flow rate

D. Poor vibration sensor installation

Note : ....

....

....

Rotating Equipment
Question 118

For typical centrifugal pumps, cavitation can result in a high vibration and an audible noise. What is a possible cause for cavitation?

A. High suction temperature (above the design limit)

B. High unbalance force (> 0.1 % of rotor weight)

C. Low shaft speed (< 70 % of full speed)

D. All of above

Note : ....

Rotating Equipment
Question 119

Consider a series of cooling water centrifugal pumps operating within a typical piping system. How does flow recirculation occur? Refer to the provided illustration.

A. Recirculation starts with the rotor's motion. As the impeller starts to rotate, the fluid in the impeller rotates (recirculates).

B. After leaving the pump (at the discharge side), the fluid goes back to the suction. Usually this happens through a pipe connecting discharge side to suction side.

C. Recirculation occurs due to the physics of vibration. Vibration leads to recirculation.

D. Recirculation develops as a result of opposing forces and moments across a coupling when shafts are severely misaligned.

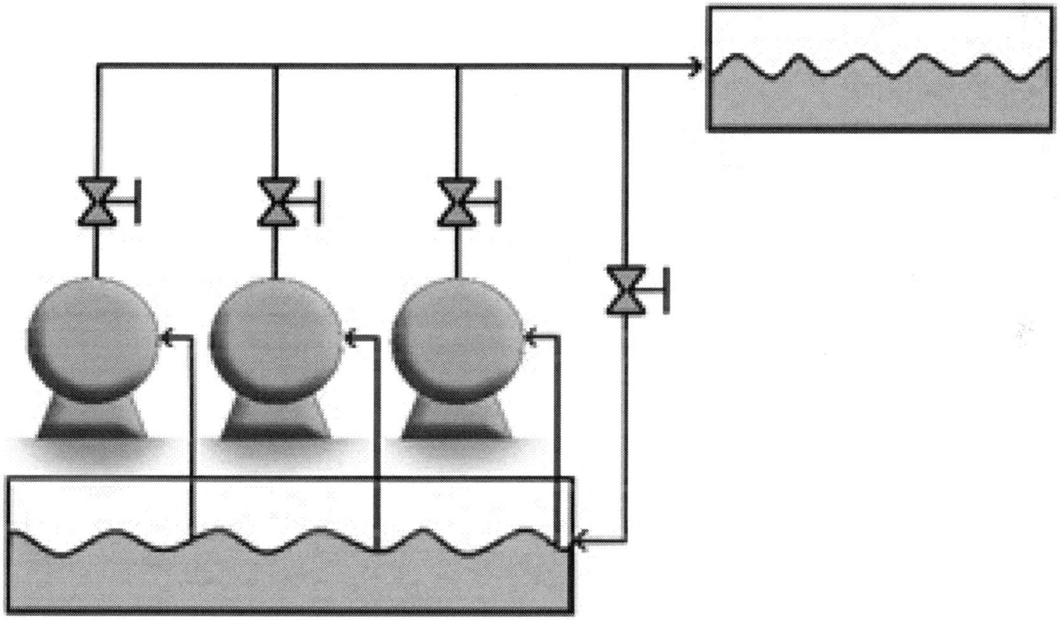

Rotating Equipment

Question 120

Why/when does a flow recirculation take place inside a centrifugal pump?

A. If pressure downstream the pump is too high compared to the pump's design, the fluid can move backwards to the pump.

B. If pressure downstream the pump is too low compared to the pump's design, the fluid moves backwards to the pump.

C. If suction pressure is too high compared to the pump's design, cavitation takes place and leads to recirculation.

D. If the pump runs exactly at the best efficiency point, flow gets unstable and eventually recirculation develops.

Rotating Equipment
Question 121

Assume industrial fans and compressors running on a closed loop process system. These fans and compressors can experience a dangerous condition if the flow rate passing through them (from suction to discharge) is too low. What is the technical term commonly used for this condition?

A. Specific Speed

B. Critical Speed

C. Surge

D. Cavitation

Note : ....

....

....

Rotating Equipment
Question 122

Study the simplified performance curve of a typical centrifugal pump. Which point (in the curve) is the best to indicate to the shut-off point? Hint: H stands for head. Q stands for flow rate.

A. 1

B. 2

C. 3

D. 4

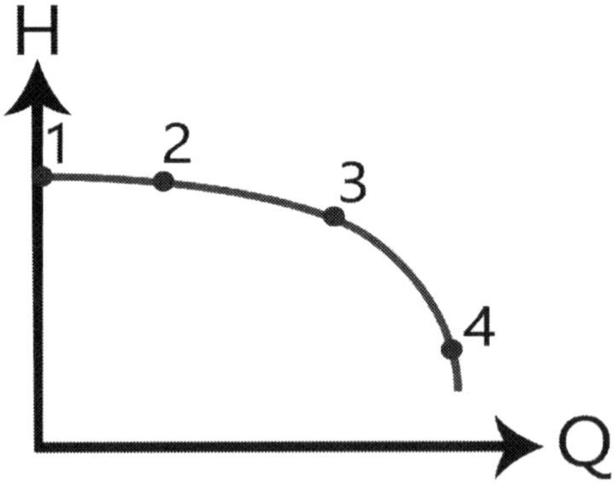

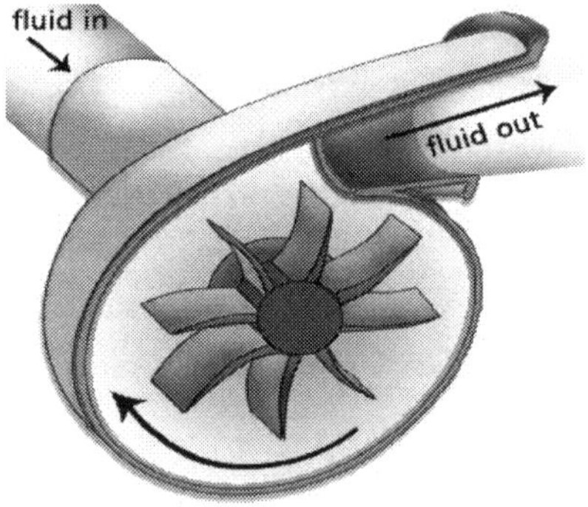

Note : ....

....

....

Rotating Equipment
Question 123

Study the simplified performance curve of a typical centrifugal pump. Which point (in the curve) is the best to indicate to the highest tendency for cavitation? Hint: H stands for head. Q stands for flow rate.

A. 1

B. 2

C. 3

D. 4

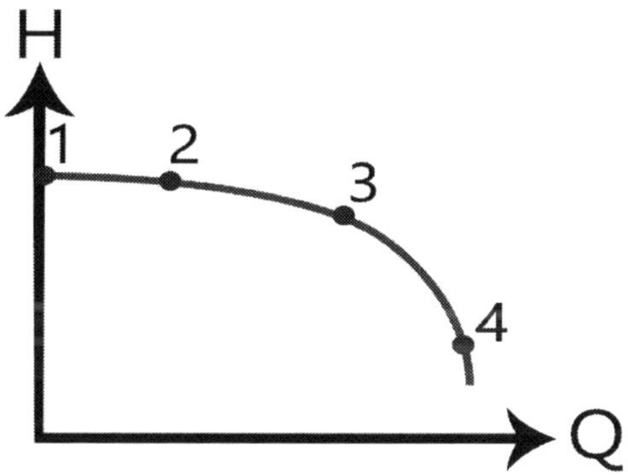

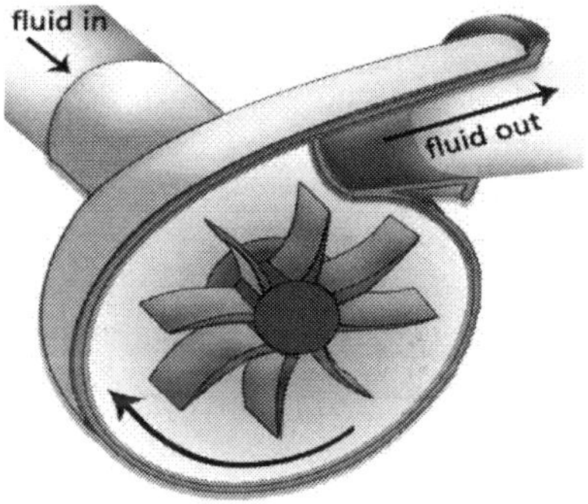

Note : ....

....

....

Rotating Equipment

Question 124

Study the simplified performance curve of a typical centrifugal pump. Which point (in the curve) is the best to possibly indicate to highest tendency for recirculation at the discharge side of the pump? Hint: H stands for Head. Q stands for flow rate.

A. 1

B. 2

C. 3

D. 4

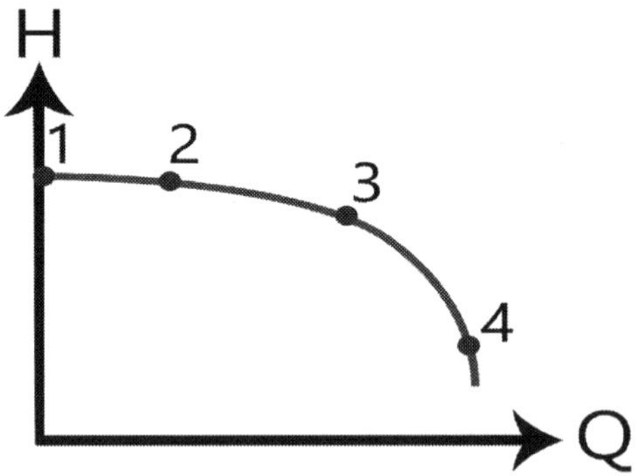

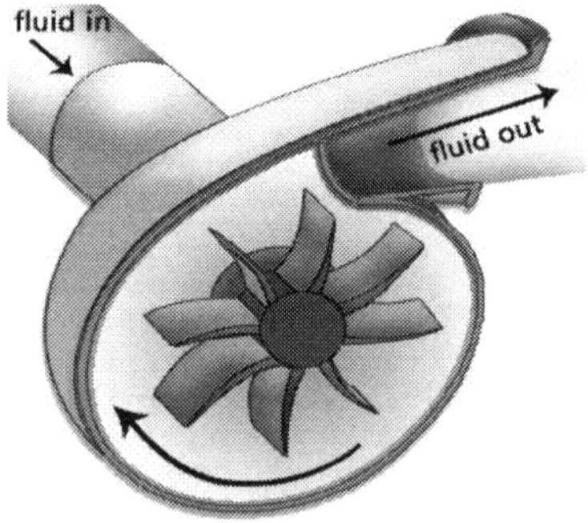

Note : ....

....

....

Rotating Equipment
Question 125

Study the simplified performance curve of a typical centrifugal pump. Which point (in the curve) most possibly indicates to the Best Efficiency Point (BEP)? Hint: H stands for Head. Q stands for flow rate.

A. 1

B. 2

C. 3

D. 4

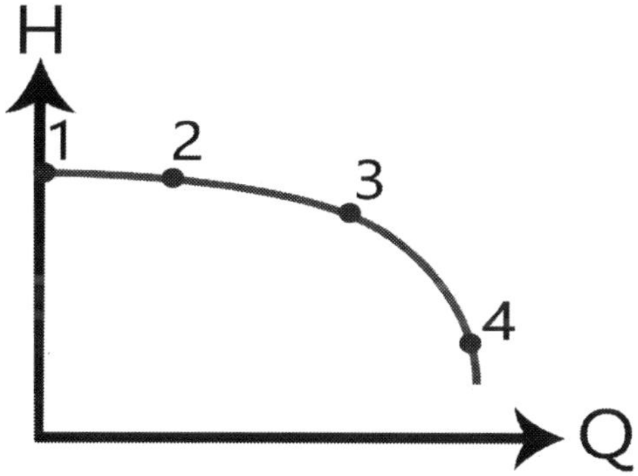

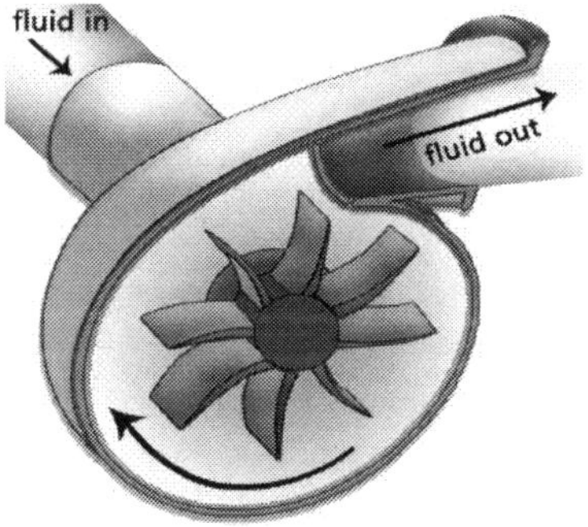

Note : ....

....

....

Rotating Equipment
Question 126

Study the below simplified characteristic curve of a typical centrifugal pump. Where does cavitation usually take place in the plot? Hint: H stands for Head. Q stands for flow rate.

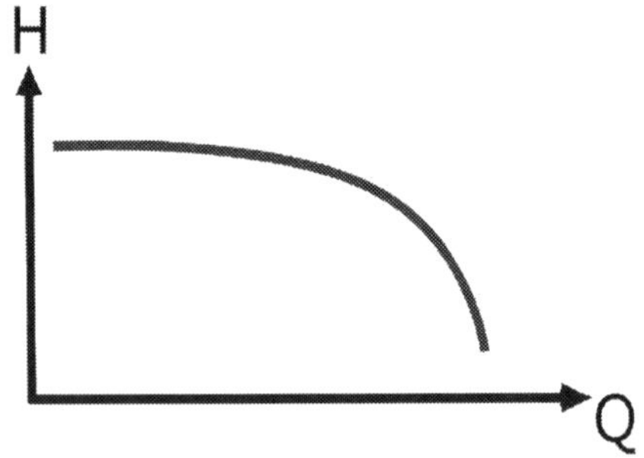

A. Identify the best efficiency point. Usually cavitation takes place to the right of the BEP.

B. Identify the best efficiency point. Usually cavitation takes place between the shut-off point and BEP.

C. Identify the shut-off point. Usually cavitation takes place to the left of the shut off point.

D. Identify the point of minimum reported flow. Usually cavitation takes place 10-20% to the right of this point.

Rotating Equipment
Question 127

How does cavitation occur when operating a typical centrifugal pump?

A. Pumps need a minimum suction pressure to operate properly. If suction pressure drops below that minimum value, cavitation occurs.

B. Pumps need a maximum suction pressure to operate properly. If suction pressure exceeds that maximum value, cavitation occurs.

C. Pump cavitation occurs if the rolling element bearings are over-lubricated.

D. Pumps are designed to operate with a certain internal clearance. If these clearances widely open, cavitation starts.

Note : ....

Rotating Equipment
Question 128

Consider a classic horizontal medium size centrifugal compressor. Which side has a higher pressure during normal operation: (inlet or outlet)? Which side has a higher temperature during normal operation: (inlet or outlet)?

	Higher Pressure	Higher Temperature
A.	Inlet	Inlet
B.	Outlet	Outlet
C.	They are equal	Outlet
D.	The higher pressure side varies depending on the flow rate	Inlet

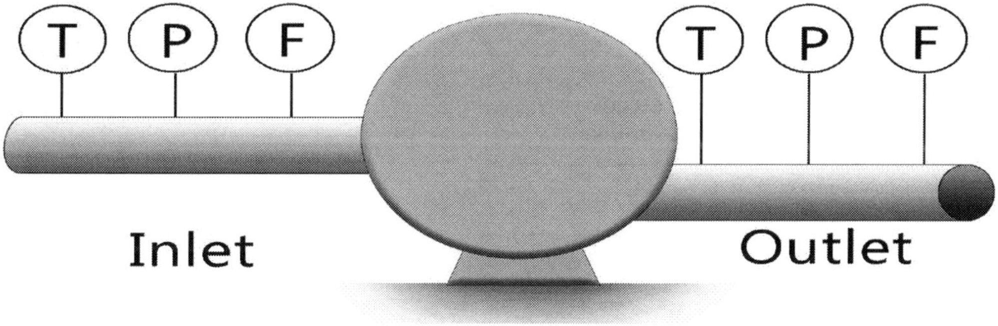
T
P
F
T
P
F
Inlet
Outlet

Rotating Equipment
Question 129

Consider a classic horizontal medium size steam turbine. Which side has a higher pressure during normal operation: (inlet or outlet)? Which side has a higher temperature during normal operation: (inlet or outlet)?

	Higher Pressure	Higher Temperature
A.	Inlet	Inlet
B.	Outlet	Outlet
C.	They are equal	Outlet
D.	The higher pressure side varies depending on the flow rate	Inlet

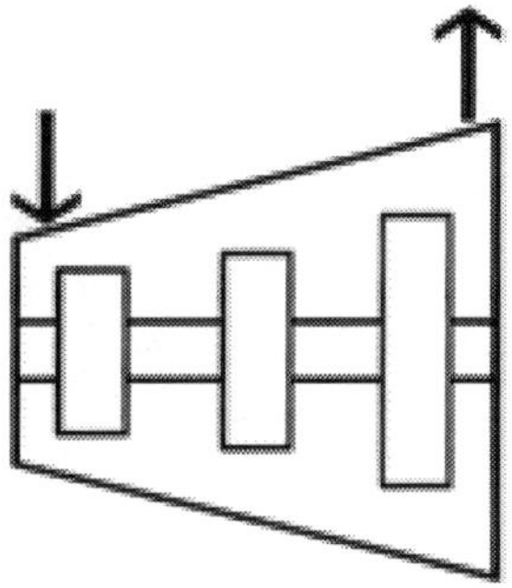

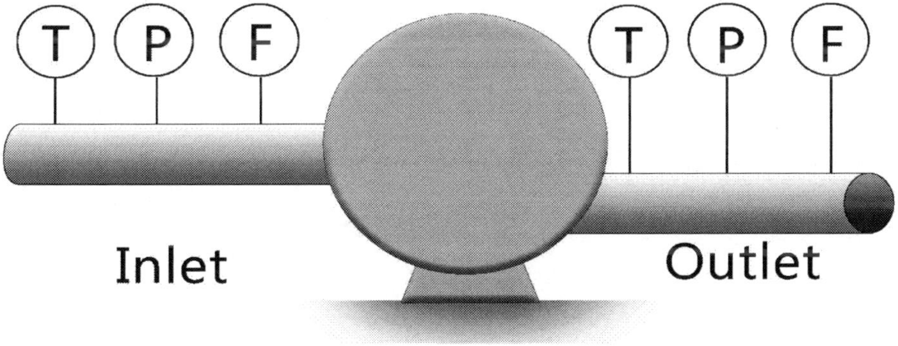
T
P
F
T
P
F
Inlet
Outlet

Rotating Equipment
Question 130

Consider a classic small size reciprocating compressor. Which side has a higher pressure and temperature during normal operation: suction or discharge?

	Higher Pressure	Higher Temperature
A.	Suction	Suction
B.	Discharge	Discharge
C.	They are equal.	Discharge
D.	The higher pressure side varies depending on the flow rate.	Suction

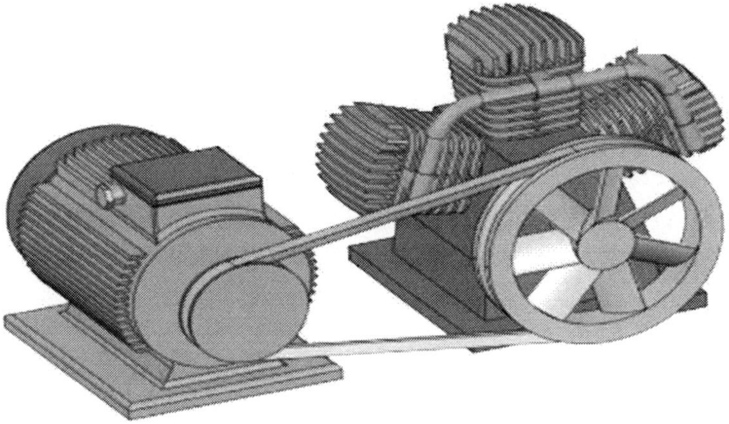

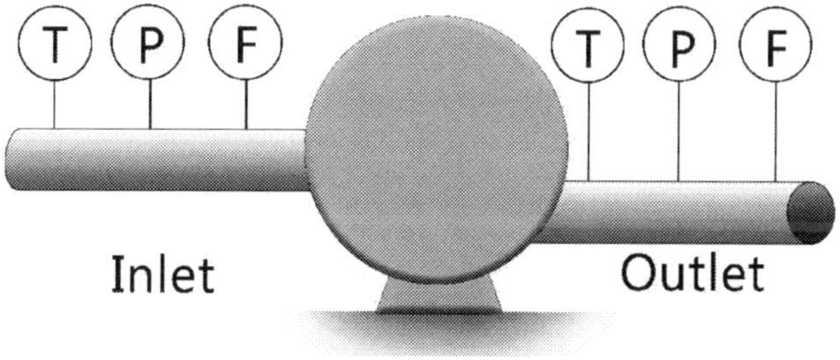
T
P
F
T
P
F
Inlet
Outlet

Rotating Equipment
Question 131

Consider a classic horizontal medium size centrifugal pump. Which side has more pressure and flow rate during normal operation: suction or discharge?

	Higher Pressure	Higher Flow
A.	Suction	Suction
B.	Discharge	They are equal.
C.	Discharge	Discharge
D.	The higher pressure side varies depending on the flow rate.	Suction

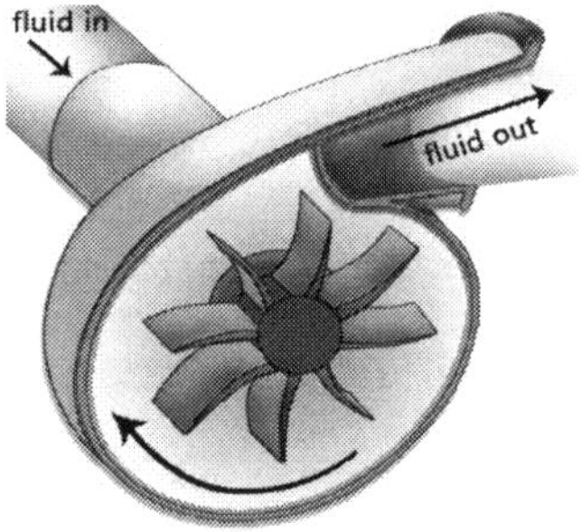

Note : ....

....

....

Rotating Equipment
Question 132

Consider a classic horizontal medium size centrifugal pump. Which location is more likely to encounter cavitation and recirculation? Hint: Refer to the next sketch a pump.

	Cavitation	Recirculation
A.	Suction	Discharge
B.	Discharge	Suction
C.	Tip of the blades	Tip of the blades
D.	At case split line	At case split line

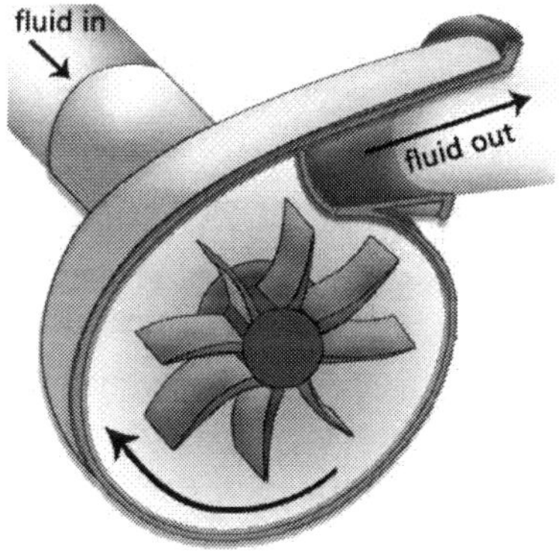

Note : ....

Rotating Equipment

Question 133

What is true about a classic gearbox and coupling?

A.	Gearbox is used to change the shaft's speed, direction or torque.
B.	Gearboxes and couplings are primarily used to increase pressure.
C.	A gearbox is primarily used to produce electric power.
D.	A coupling is primarily used to amplify rotational power.

Note : ....

....

....

Rotating Equipment
Question 134

What component of rotating equipment is shown in the next illustration? What is its function?

	Part	Function
A.	Rubber coupling	Power transmission
B.	Rolling element bearing	To hold shaft
C.	Gearbox	Power storage
D.	Bearing housing	To control leakage

Rotating Equipment
Question 135

Which of the following are four common types of gears that are used in the modern industry?

A.	Ball	Roller	Sphere	Needle
B.	Jawless	2 Jaw	3 Jaw	4 Jaw
C.	Forward	Backward	Rigid	Flexible
D.	Worm	Spur	Helical	Bevel

Note : ....

....

....

Rotating Equipment
Question 136

What type of equipment is shown in the next images? Hint: it is used with large machinery. The upper half of the case is removed.

A. Gearbox

B. Centrifugal pump

C. Gas turbine

D. Forced draft fan

Note : ....

....

....

Rotating Equipment
Question 137

What is true about a classic gearbox and coupling? Refer to the simplified images for a gearbox and coupling. Select the best option from the given answers.

	Gears are_________. 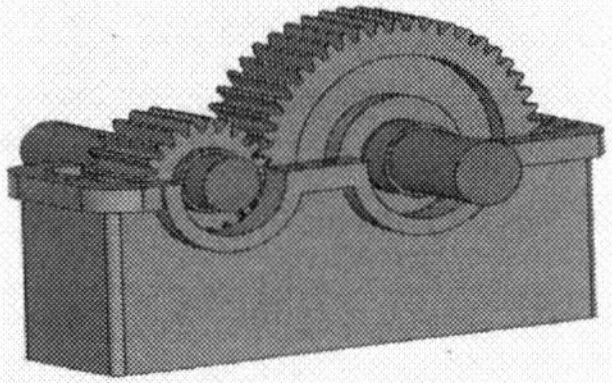	Couplings are___________. 
A.	used to transmit rotational power	used to transmit rotational power
B.	used to increase rotational speed.	used to increase rotational speed.
C.	used only with low speeds.	used low and high speeds.
D.	used only with high speeds.	used low and high speeds.

Rotating Equipment
Question 138

What is the purpose of installing a coupling for a motor driving a pump?

A. It prevents the translational motions (axial vibrations).

B. It transmits the rotational power from the driver to the driven equipment.

C. It prevents the shaft to rotate in the wrong direction of rotation.

D. It reduces thermal growth and shims are added to adjust the final shaft alignment.

Note : ....

....

....

Rotating Equipment
Question 139

Which of the following shows correct description for the adjacent illustration?

A.	Only 1 and 2
B.	Only 2 and 3
C.	Only 3 and 4
D.	All of the above

TYPES OF GEARS

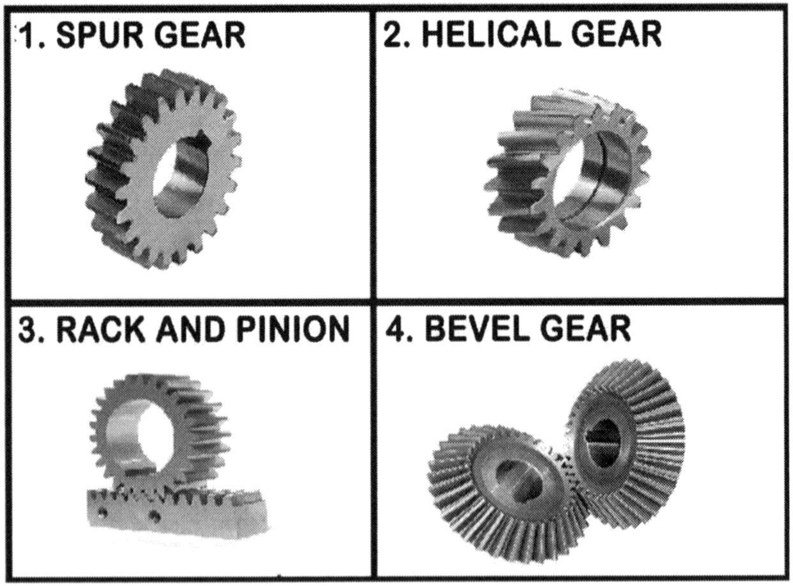

Rotating Equipment
Question 140

Which of the following are helical gears?

A.	Only 1
B.	Only 2
C.	Only 3
D.	All of the above

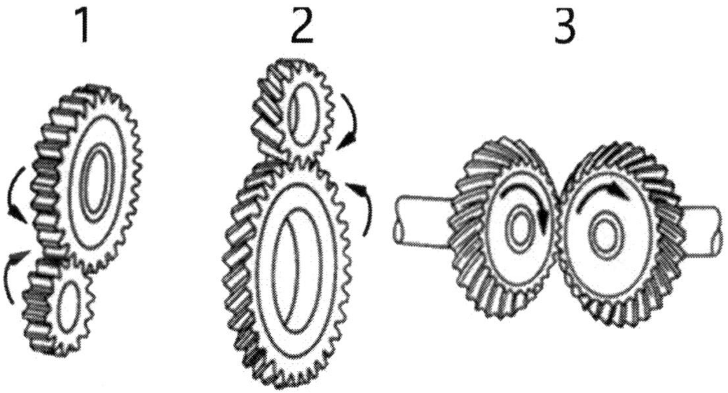

Note : ....

....

....

Rotating Equipment
Question 141

What is the difference between a single –stage speed reduction gearbox and a double -stage speed reduction gearbox?

	Single-Stage	Double-Stage
A.	The shafts are parallel to each other.	The shafts are perpendicular to each other.
B.	The teeth are made of normal alloys.	The teeth are made of very hard alloys.
C.	The number of teeth in both input and output gears are different.	The number of teeth in both input and output gears are equal.
D.	The speed ratio is (usually) lower and directions of rotation (for input and output shafts) are opposite.	The speed ratio is (usually) higher and directions of rotation (for both input and output shafts) are identical.

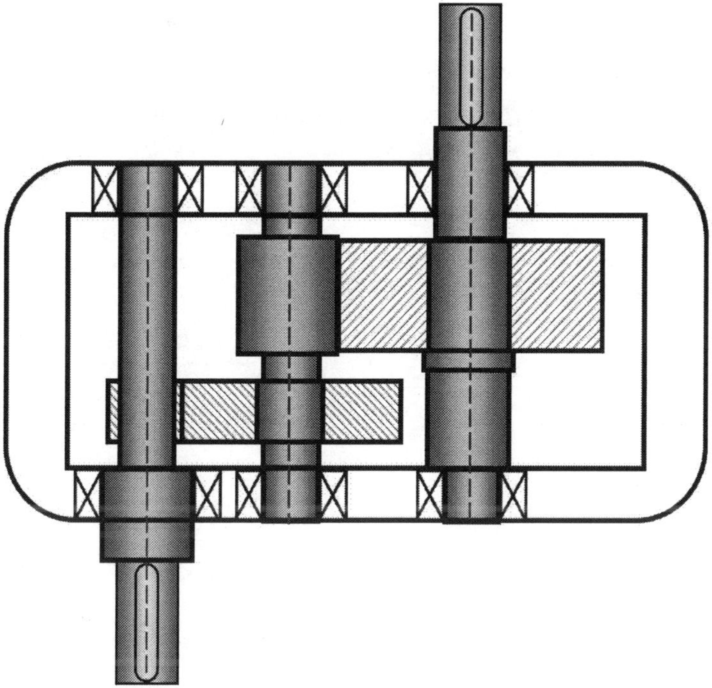

Note :

Rotating Equipment
Question 142

Refer to the next illustration for a typical simple gearbox. How many radial bearings are used to carry the shafts?

A. 2 bearings

B. 4 bearings

C. 6 bearings

D. 8 bearings

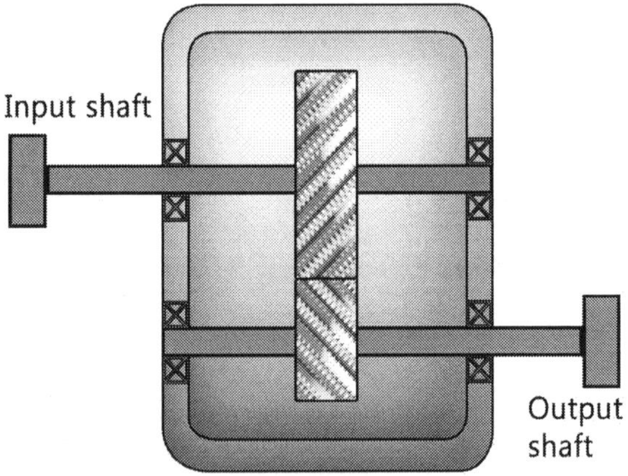

Rotating Equipment
Question 143

The next illustration is for a motor (1780 rpm) driving a vertical shaft of a cooling fan through a gearbox. In addition to transmitting the power from the motor to the fan, what is a key function this gearbox provides?

A. Changing the orientation of the shaft centerline (from horizontal to vertical).

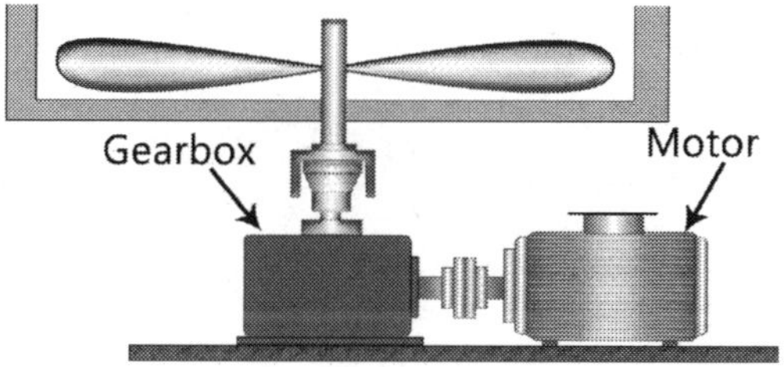

B. Changing the direction of rotation (from CW to CCW or vice versa).

C. Simultaneously increasing both torque and speed (from the motor to the fan).

D. Changing the frequency from Hz to cpm.

Note : ....

Rotating Equipment
Question 144

Consider a simple single stage speed increasing gearbox. Assume the power transmitted is < 100 HP (75 kw) and input speed is 1790 rpm. What is the direction of rotation of the output shaft?

A. Clockwise

B. Counter Clockwise

C. Same direction of rotation of the that of input shaft

D. Opposite direction of rotation to the input shaft

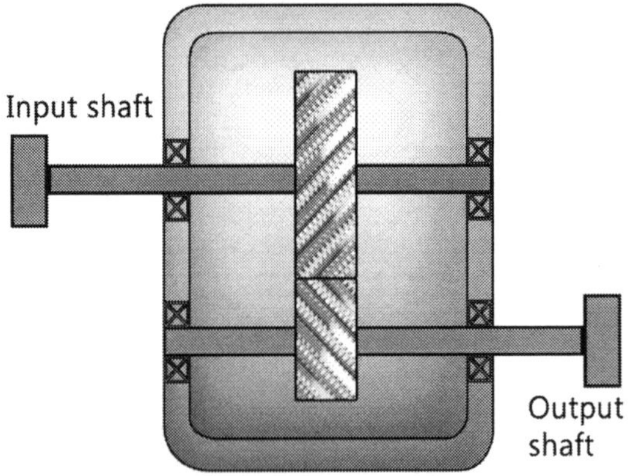

Note : ....

....

....

Rotating Equipment
Question 145

Study the shown unit of a motor driving a compressor through a gear box. Which side of the gearbox is referred to as output shaft?

A. The low speed shaft

B. The high speed shaft

C. The shaft with the higher power

D. The shaft connected to the motor

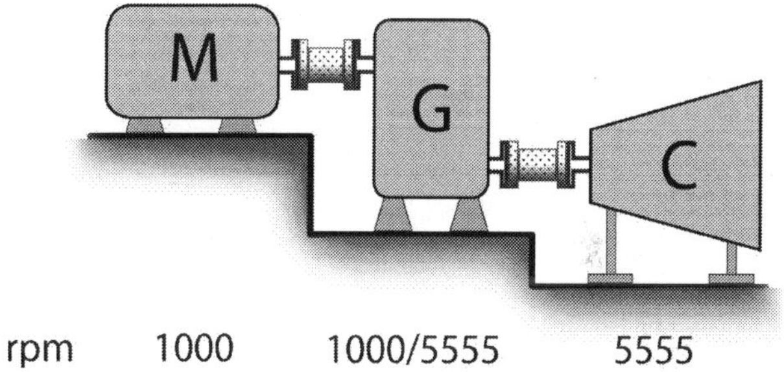

Rotating Equipment
Question 146

For a typical gearbox, which side of the gearbox is referred to as the input shaft? Which side of the gearbox is referred to as the output shaft? Refer to the illustration.

	Input	Output
A.	The shaft of the high speed	The shaft of the low speed
B.	The shaft connected to the driver equipment	The shaft connected to the driven equipment
C.	The shaft connected to the driven machine	The shaft connected to the driven machine
D.	The shaft of the pinion gear	The shaft of the pinion gear

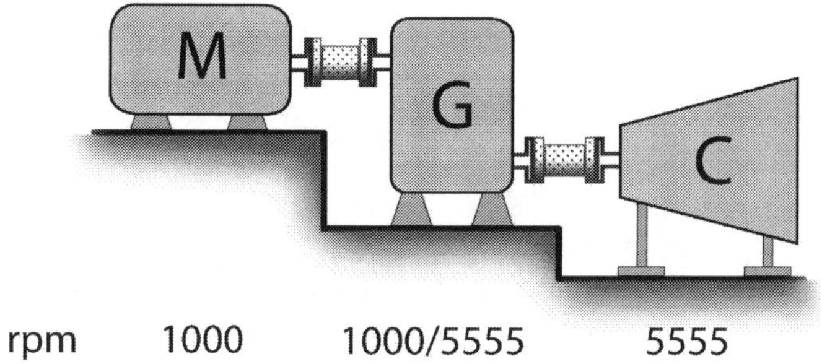
M
G
C
rpm
1000
1000/5555
5555

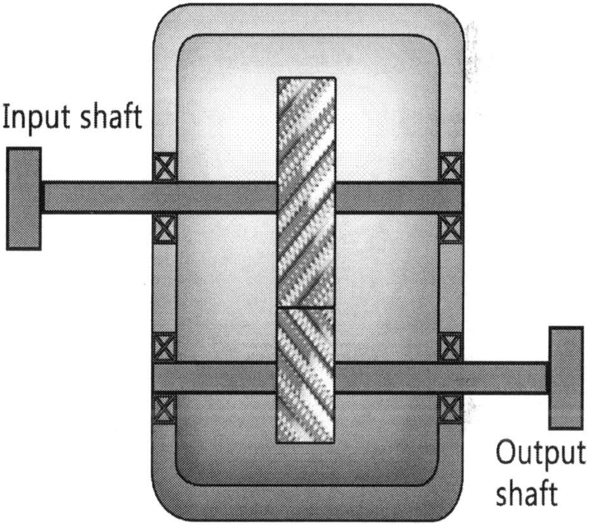
Input shaft
Output shaft

Rotating Equipment
Question 147

In the modern industry, all gearboxes are designed to increase the speed of the driven equipment. True or False?

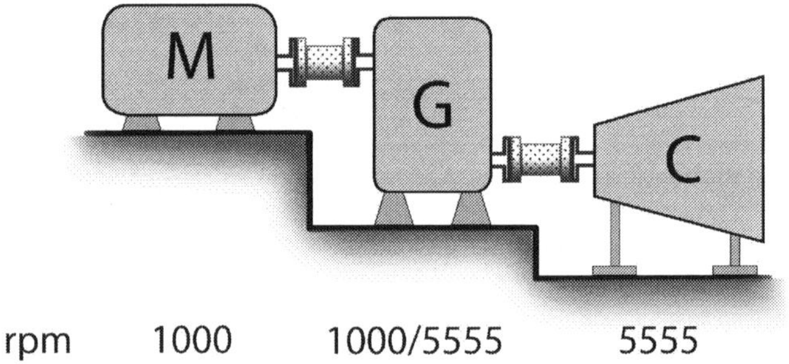

A. True, because common AC motors have a maximum speed of 3600 rpm.

B. True, gearboxes are used for high power machines which always require a high speed.

C. False. Some gearboxes are used to reduce speed.

D. False. In fact, input speed must be higher than the output speed.

Rotating Equipment
Question 148

What is the best description for the gears in the shown gearbox?

A.	Worm gear
B.	Spur gear
C.	Helical gear
D.	Bevel gear

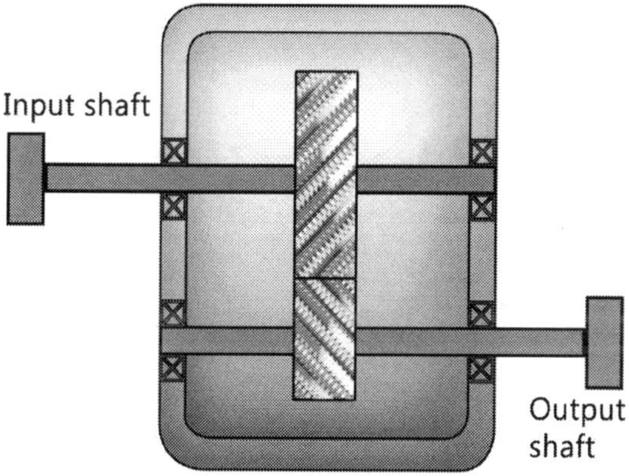

Note : ....

....

....

Rotating Equipment
Question 149

Refer to the next simplified illustration for a gearbox. What are the parts indicated by numbers 1 and 2?

	1	2
A.	Pinion	High speed gear
B.	Gear	Pinion
C.	Low speed gear	Gear
D.	Thrust bearing	Thrust disk

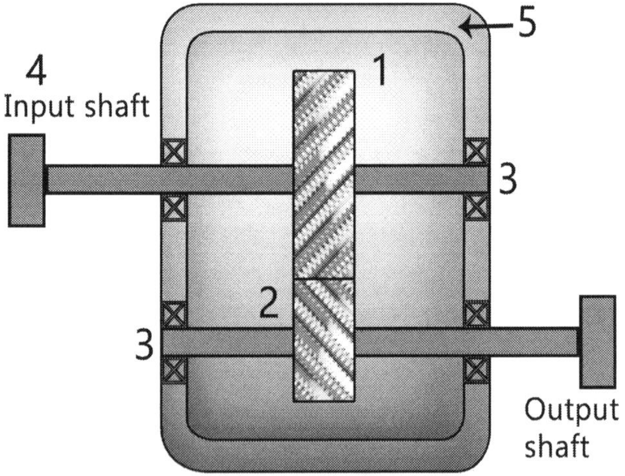

Note :

Rotating Equipment
Question 150

Refer to the next simplified illustration for a gearbox. What are the parts indicated by numbers 4 and 5?

	4	5
A.	High speed shaft	Case
B.	Thrust disk	Pinion
C.	Low speed Shaft	Housing
D.	Diffuser	Outer race

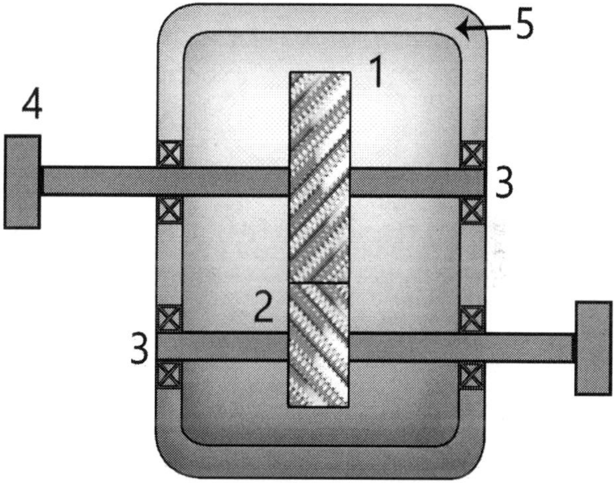

Note : ....

Rotating Equipment
Question 151

What is true about all gearboxes in general?

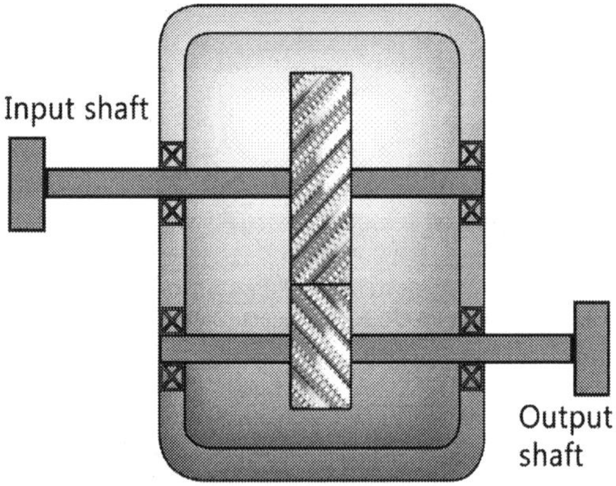

A. The speed of input shaft equals that of the output shaft.

B. The direction of rotation of output shaft is opposite to that of the input shaft.

C. The centerline of input shaft is perpendicular to the output shaft.

D. The power of input shaft equals that of the output shaft (assuming no power losses).

Rotating Equipment
Question 152

What is a key distinct feature of bevel gears compared to other types of gearboxes? Hint: Sketch this gear type.

A. Speed of input shaft equals that of the output shaft.

B. The gear mesh frequency is not constant, even if the shafts run at constant speeds.

C. The centerline of output shaft is perpendicular to that of the input shaft.

D. The power of input shaft equals that of the output shaft (assuming no power loss).

Note : ....

....

....

Rotating Equipment
Question 153

In a basic gearbox design, what is the meaning of "Common Factor"? Give an example.

A. It is an integer number that can divide the number of teeth in both pinion and gear (without leaving a remainder). Example: 3 is a common factor for 33 teeth and 90 teeth.

B. The average speed of the input and output shafts. Example: 1500 cpm is the common factor for 1000 and 2000 rpm.

C. Another term for the gear mesh frequency which equals speed x teeth. Example: 360000 cpm is the common factor for 3600 rpm with 100 teeth.

D. It is gear mesh frequency quality (GMFQ) which is calculated by = teeth (gear) x teeth (pinion). Example: 2700 for 30 teeth and 90 teeth.

Rotating Equipment
Question 154

What is the gear mesh frequency for the following simple gearbox? Input shaft has 23 teeth and turns at 1800 rpm while its output shaft has 57 teeth. What is the speed of the output shaft? Hint: Add the numbers to the illustration.

	GMF	Output Speed
A.	690 cpm	4460 rpm
B.	690 Hz	726 rpm
C.	1311 Hz	690 cpm
D.	41400 Hz	Cannot be found from the given information

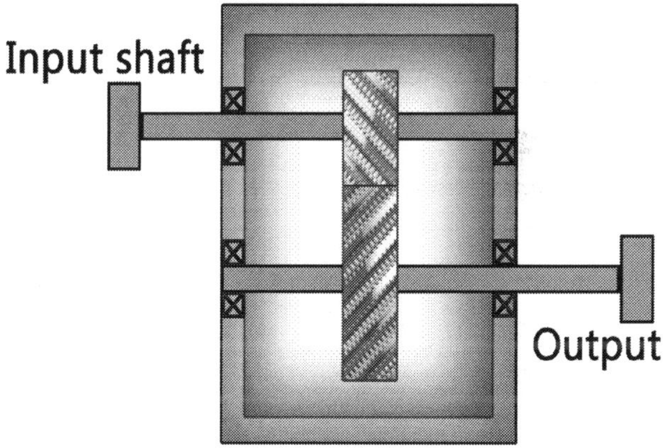

Rotating Equipment
Question 155

What is the gear mesh frequency for a simple gearbox whose pinion has 27 teeth and turns at 900 rpm? What is the speed of the output shaft? Hint: Add the numbers to the illustration.

	GMF	Speed
A.	33.3 Hz	2000 Hz
B.	33.3 Hz	Cannot be found from the given information
C.	405 Hz	Cannot be found from the given information
D.	24300 cpm	405 Hz

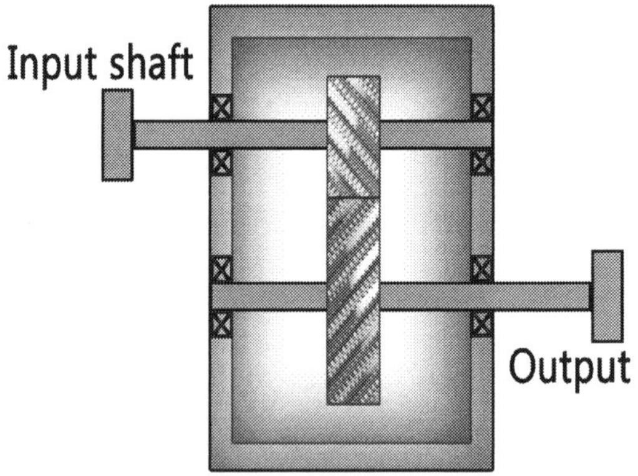

Rotating Equipment
Question 156

The next illustration is for a motor driving a cooling fan above a heat exchanger. What is the direction of rotation of the fan? The motor turns clockwise when one looks at it from the NDE.

A. Clockwise.

B. Counter clockwise.

C. More data is needed to answer.

D. It depends on the running speed and critical speed.

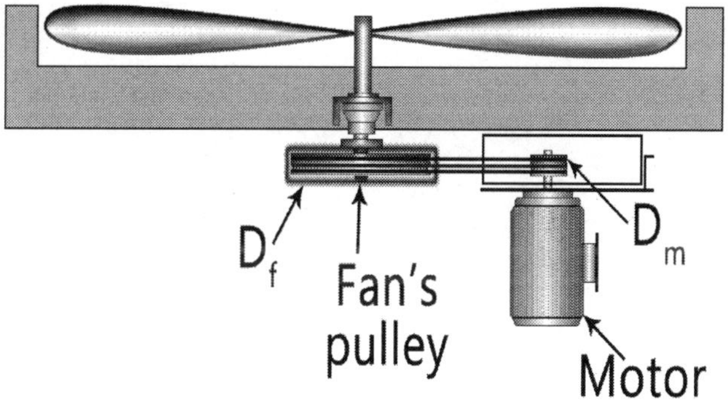

Note :

Rotating Equipment
Question 157

The next illustration is for a motor driving a cooling fan. What is true about the speed of the fan? The motor turns clockwise when one looks at it from the NDE.

A. It is faster than the motor's speed.

B. It is slower than the motor's speed.

C. It is equal to the motor's speed.

D. It depends on the critical speed.

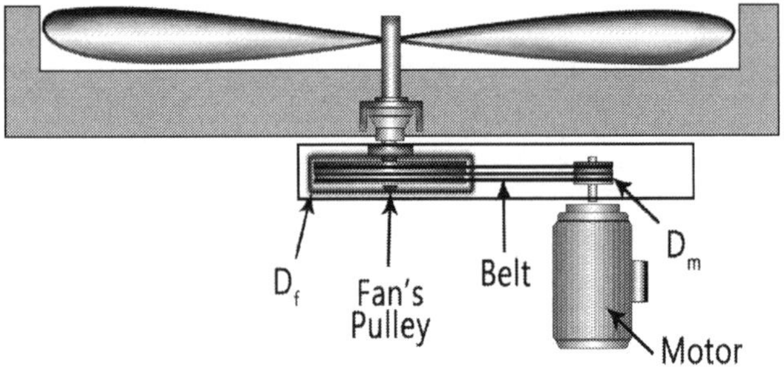

Note : ....

....

....

Rotating Equipment
Question 158

Using the next illustration, estimate the speed of the driven shaft (fan) if the driver induction motor runs at 1795 rpm. The diameter of motor's pulley D_m = 15 inch while the other pulley is D_f = 45 inch. The motor is rated at 50 HP but the fan is rated at 40 HP. Expect some rounding.

A. 360 rpm

B. 514 rpm

C. 600 rpm

D. 90 Hz

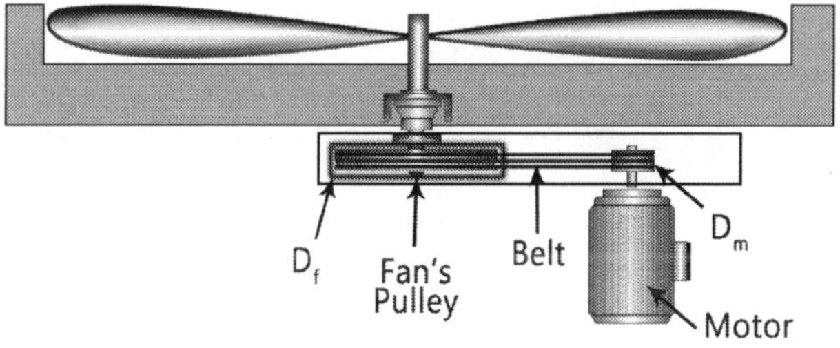

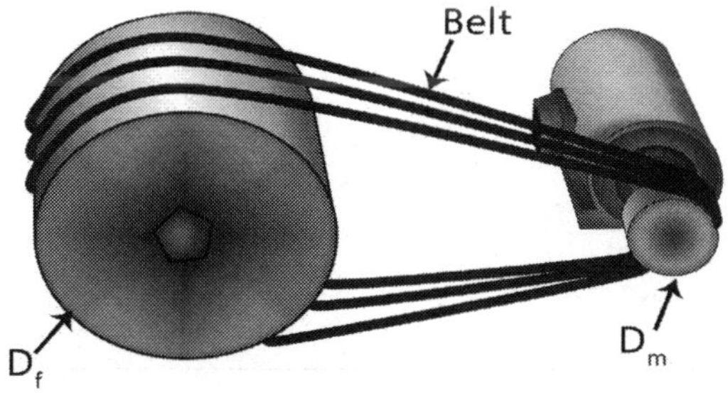

Rotating Equipment
Question 159

Using the next illustration, estimate the speed of the driven shaft (fan) if the driver induction motor runs at 1495 rpm. The diameter of motor's pulley D_m = 30 cm while the other pulley is D_f = 90 cm. The motor is rated at 50 kW but the fan is rated at 40 kW. Expect some rounding.

A. 360 rpm

B. 500 rpm

C. 600 rpm

D. 90 Hz

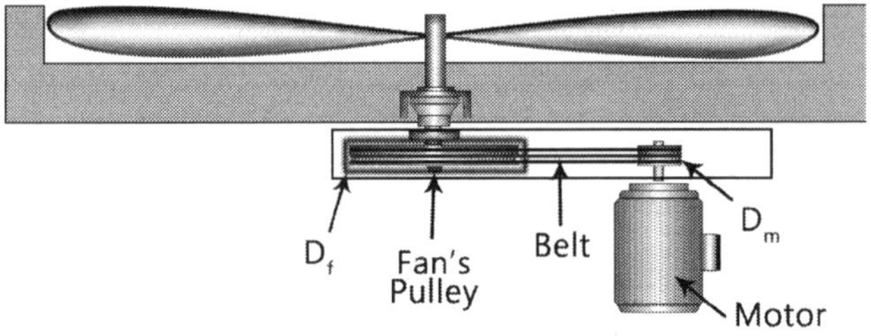

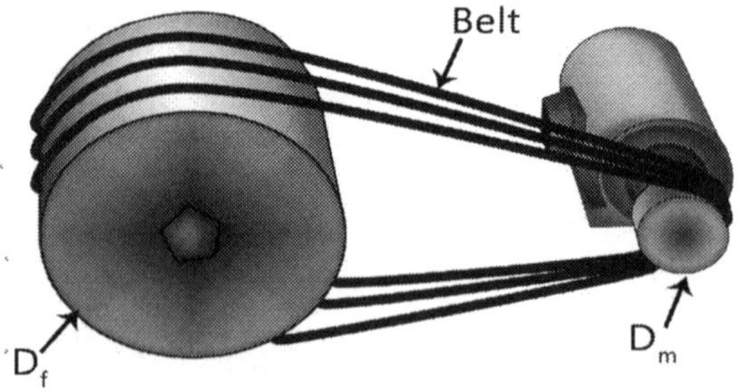

Note : ....

....

....

Rotating Equipment
Question 160

Using the next illustration, estimate the speed of the driven shaft if the motor runs at 3595 rpm. The driven pulley is 3.2 times larger than that of the motor.

A. 600 rpm

B. 1123 rpm

C. 1890 rpm

D. 3598 rpm

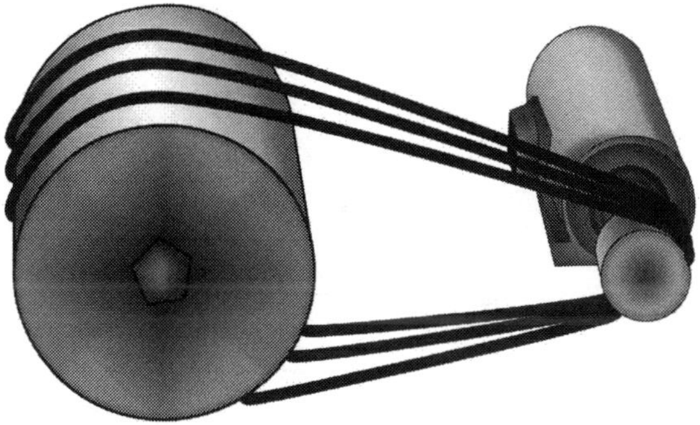

Rotating Equipment
Question 161

Using the next illustration, estimate the speed of the driven shaft if the induction motor runs at 1790 rpm. Use a ruler to estimate any missing data. Assume the image is drawn to scale.

A. 560 rpm

B. 814 rpm

C. 1810 rpm

D. 3598 rpm

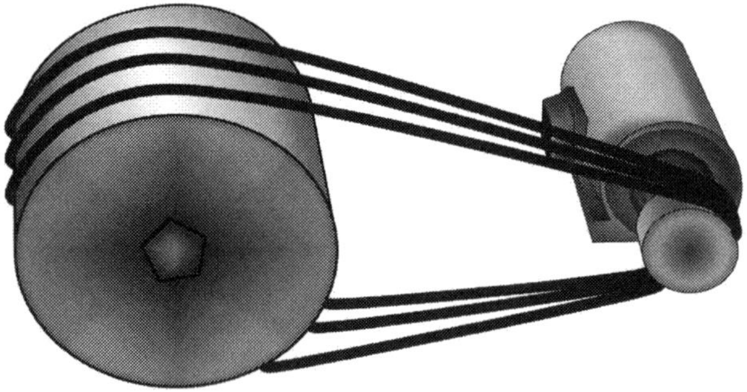

Note : ....

....

....

Rotating Equipment
Question 162

Using the next illustration, estimate the speed of the driven shaft if the motor runs at 1795 rpm. Use a ruler to estimate any missing data. The fan runs at 80% of its design capacity (or 80% of its design power). Expect some rounding

A. 357 rpm

B. 455 rpm

C. 1424 rpm

D. 6297 rpm

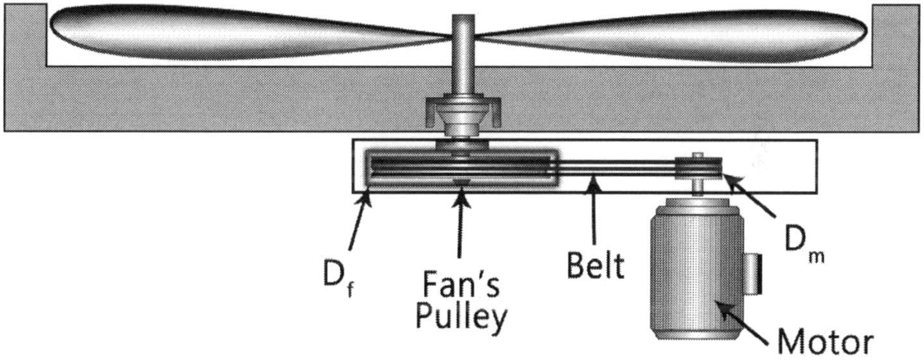

Note : ....

....

....

ANSWER SHEETS

EMPTY ANSWER SHEET 1

1	A	B	C	D
2	A	B	C	D
3	A	B	C	D
4	A	B	C	D
5	A	B	C	D
6	A	B	C	D
7	A	B	C	D
8	A	B	C	D
9	A	B	C	D
10	A	B	C	D
11	A	B	C	D
12	A	B	C	D
13	A	B	C	D
14	A	B	C	D
15	A	B	C	D

16	A	B	C	D
17	A	B	C	D
18	A	B	C	D
19	A	B	C	D
20	A	B	C	D
21	A	B	C	D
22	A	B	C	D
23	A	B	C	D
24	A	B	C	D
25	A	B	C	D
26	A	B	C	D
27	A	B	C	D
28	A	B	C	D
29	A	B	C	D
30	A	B	C	D

EMPTY ANSWER SHEET 1

31	A	B	C	D
32	A	B	C	D
33	A	B	C	D
34	A	B	C	D
35	A	B	C	D
36	A	B	C	D
37	A	B	C	D
38	A	B	C	D
39	A	B	C	D
40	A	B	C	D
41	A	B	C	D
42	A	B	C	D
43	A	B	C	D
44	A	B	C	D
45	A	B	C	D

46	A	B	C	D
47	A	B	C	D
48	A	B	C	D
49	A	B	C	D
50	A	B	C	D
51	A	B	C	D
52	A	B	C	D
53	A	B	C	D
54	A	B	C	D
55	A	B	C	D
56	A	B	C	D
57	A	B	C	D
58	A	B	C	D
59	A	B	C	D
60	A	B	C	D

EMPTY ANSWER SHEET 1

61	A	B	C	D
62	A	B	C	D
63	A	B	C	D
64	A	B	C	D
65	A	B	C	D
66	A	B	C	D
67	A	B	C	D
68	A	B	C	D
69	A	B	C	D
70	A	B	C	D
71	A	B	C	D
72	A	B	C	D
73	A	B	C	D
74	A	B	C	D
75	A	B	C	D

76	A	B	C	D
77	A	B	C	D
78	A	B	C	D
79	A	B	C	D
80	A	B	C	D
81	A	B	C	D
82	A	B	C	D
83	A	B	C	D
84	A	B	C	D
85	A	B	C	D
86	A	B	C	D
87	A	B	C	D
88	A	B	C	D
89	A	B	C	D
90	A	B	C	D

EMPTY ANSWER SHEET 1

91	A	B	C	D
92	A	B	C	D
93	A	B	C	D
94	A	B	C	D
95	A	B	C	D
96	A	B	C	D
97	A	B	C	D
98	A	B	C	D
99	A	B	C	D
100	A	B	C	D
101	A	B	C	D
102	A	B	C	D
103	A	B	C	D
104	A	B	C	D
105	A	B	C	D

106	A	B	C	D
107	A	B	C	D
108	A	B	C	D
109	A	B	C	D
110	A	B	C	D
111	A	B	C	D
112	A	B	C	D
113	A	B	C	D
114	A	B	C	D
115	A	B	C	D
116	A	B	C	D
117	A	B	C	D
118	A	B	C	D
119	A	B	C	D
120	A	B	C	D

EMPTY ANSWER SHEET 1

121	A	B	C	D
122	A	B	C	D
123	A	B	C	D
124	A	B	C	D
125	A	B	C	D
126	A	B	C	D
127	A	B	C	D
128	A	B	C	D
129	A	B	C	D
130	A	B	C	D
131	A	B	C	D
132	A	B	C	D
133	A	B	C	D
134	A	B	C	D
135	A	B	C	D

136	A	B	C	D
137	A	B	C	D
138	A	B	C	D
139	A	B	C	D
140	A	B	C	D
141	A	B	C	D
142	A	B	C	D
143	A	B	C	D
144	A	B	C	D
145	A	B	C	D
146	A	B	C	D
147	A	B	C	D
148	A	B	C	D
149	A	B	C	D
150	A	B	C	D

EMPTY ANSWER SHEET 1

151	A	B	C	D
152	A	B	C	D
153	A	B	C	D
154	A	B	C	D
155	A	B	C	D
156	A	B	C	D
157	A	B	C	D
158	A	B	C	D
159	A	B	C	D
160	A	B	C	D
161	A	B	C	D
162	A	B	C	D

Notes

EMPTY ANSWER SHEET 2

1	A	B	C	D
2	A	B	C	D
3	A	B	C	D
4	A	B	C	D
5	A	B	C	D
6	A	B	C	D
7	A	B	C	D
8	A	B	C	D
9	A	B	C	D
10	A	B	C	D
11	A	B	C	D
12	A	B	C	D
13	A	B	C	D
14	A	B	C	D
15	A	B	C	D

16	A	B	C	D
17	A	B	C	D
18	A	B	C	D
19	A	B	C	D
20	A	B	C	D
21	A	B	C	D
22	A	B	C	D
23	A	B	C	D
24	A	B	C	D
25	A	B	C	D
26	A	B	C	D
27	A	B	C	D
28	A	B	C	D
29	A	B	C	D
30	A	B	C	D

EMPTY ANSWER SHEET 2

31	A	B	C	D
32	A	B	C	D
33	A	B	C	D
34	A	B	C	D
35	A	B	C	D
36	A	B	C	D
37	A	B	C	D
38	A	B	C	D
39	A	B	C	D
40	A	B	C	D
41	A	B	C	D
42	A	B	C	D
43	A	B	C	D
44	A	B	C	D
45	A	B	C	D

46	A	B	C	D
47	A	B	C	D
48	A	B	C	D
49	A	B	C	D
50	A	B	C	D
51	A	B	C	D
52	A	B	C	D
53	A	B	C	D
54	A	B	C	D
55	A	B	C	D
56	A	B	C	D
57	A	B	C	D
58	A	B	C	D
59	A	B	C	D
60	A	B	C	D

EMPTY ANSWER SHEET 2

61	A	B	C	D
62	A	B	C	D
63	A	B	C	D
64	A	B	C	D
65	A	B	C	D
66	A	B	C	D
67	A	B	C	D
68	A	B	C	D
69	A	B	C	D
70	A	B	C	D
71	A	B	C	D
72	A	B	C	D
73	A	B	C	D
74	A	B	C	D
75	A	B	C	D

76	A	B	C	D
77	A	B	C	D
78	A	B	C	D
79	A	B	C	D
80	A	B	C	D
81	A	B	C	D
82	A	B	C	D
83	A	B	C	D
84	A	B	C	D
85	A	B	C	D
86	A	B	C	D
87	A	B	C	D
88	A	B	C	D
89	A	B	C	D
90	A	B	C	D

EMPTY ANSWER SHEET 2

91	A	B	C	D
92	A	B	C	D
93	A	B	C	D
94	A	B	C	D
95	A	B	C	D
96	A	B	C	D
97	A	B	C	D
98	A	B	C	D
99	A	B	C	D
100	A	B	C	D
101	A	B	C	D
102	A	B	C	D
103	A	B	C	D
104	A	B	C	D
105	A	B	C	D

106	A	B	C	D
107	A	B	C	D
108	A	B	C	D
109	A	B	C	D
110	A	B	C	D
111	A	B	C	D
112	A	B	C	D
113	A	B	C	D
114	A	B	C	D
115	A	B	C	D
116	A	B	C	D
117	A	B	C	D
118	A	B	C	D
119	A	B	C	D
120	A	B	C	D

EMPTY ANSWER SHEET 2

121	A	B	C	D
122	A	B	C	D
123	A	B	C	D
124	A	B	C	D
125	A	B	C	D
126	A	B	C	D
127	A	B	C	D
128	A	B	C	D
129	A	B	C	D
130	A	B	C	D
131	A	B	C	D
132	A	B	C	D
133	A	B	C	D
134	A	B	C	D
135	A	B	C	D

136	A	B	C	D
137	A	B	C	D
138	A	B	C	D
139	A	B	C	D
140	A	B	C	D
141	A	B	C	D
142	A	B	C	D
143	A	B	C	D
144	A	B	C	D
145	A	B	C	D
146	A	B	C	D
147	A	B	C	D
148	A	B	C	D
149	A	B	C	D
150	A	B	C	D

EMPTY ANSWER SHEET 2

151	A	B	C	D
152	A	B	C	D
153	A	B	C	D
154	A	B	C	D
155	A	B	C	D
156	A	B	C	D
157	A	B	C	D
158	A	B	C	D
159	A	B	C	D
160	A	B	C	D
161	A	B	C	D
162	A	B	C	D

Notes

Notes

ANSWER KEY

Q. #	Answer
1	A
2	A
3	D
4	A
5	A
6	C
7	C
8	A
9	C
10	B
11	B
12	A
13	A
14	A
15	A

Q. #	Answer
16	B
17	D
18	A
19	D
20	D
21	D
22	A
23	B
24	A
25	D
26	D
27	C
28	D
29	A
30	D

Q. #	Answer
31	D
32	D
33	B
34	A
35	A
36	D
37	A
38	D
39	A
40	D
41	B
42	B
43	B
44	D
45	C

Q. #	Answer
46	B
47	D
48	C
49	D
50	D
51	D
52	D
53	A
54	D
55	B
56	C
57	B
58	A
59	C
60	C

ANSWER KEY

Q. #	Answer
61	A
62	B
63	A
64	C
65	C
66	B
67	A
68	B
69	D
70	B
71	D
72	A
73	A
74	D
75	A

Q. #	Answer
76	A
77	D
78	D
79	B
80	B
81	B
82	B
83	B
84	A
85	A
86	B
87	C
88	A
89	B
90	D

Q. #	Answer
91	C
92	D
93	A
94	D
95	D
96	D
97	D
98	C
99	C
100	A
101	B
102	A
103	C
104	B
105	A

Q. #	Answer
106	A
107	B
108	A
109	D
110	B
111	B
112	D
113	A
114	A
115	C
116	B
117	A
118	A
119	B
120	A

ANSWER KEY

Q. #	Answer
121	C
122	A
123	D
124	A
125	C
126	A
127	A
128	B
129	A
130	B
131	B
132	A
133	A
134	A
135	D

Q. #	Answer
136	A
137	A
138	B
139	D
140	B
141	D
142	B
143	A
144	D
145	B
146	B
147	C
148	C
149	B
150	C

Q. #	Answer
151	D
152	C
153	A
154	B
155	C
156	A
157	B
158	C
159	B
160	B
161	A
162	B

Notes

ORDER OTHER PARTS OF CAT I PREP I PACKAGE

Don't guess where your skill stands; certify it. PrepCertify believes that the best preparation for professional certifications is obtained through practicing well-designed real world problems.

Learn what really matters in real world industry while mastering the Body of Knowledge in the certification standards. Your Cat I Prep I series does that for you. Through PrepCertify, you will achieve your certification in a much shorter time and with a greater value of your time and effort.

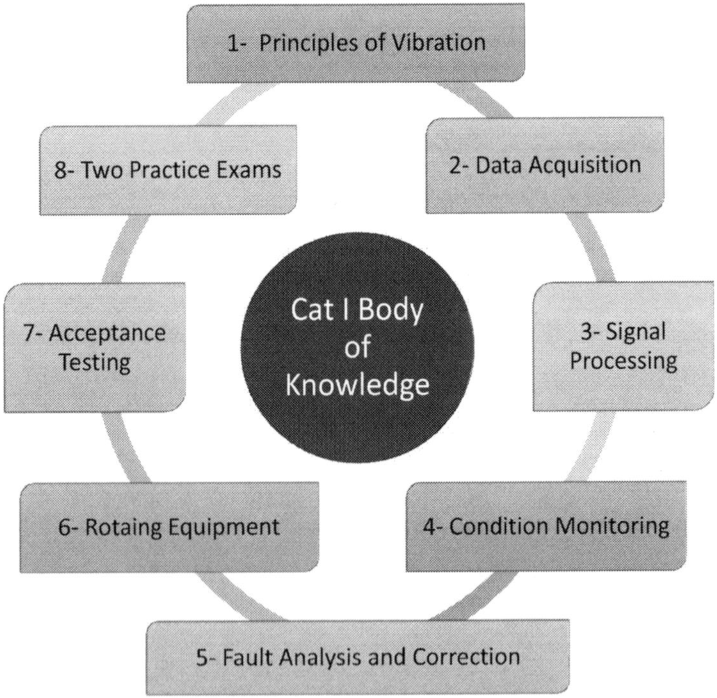

Vibration Analysis Certification Exam Preparation Package
Certified Vibration Analyst Category I
ISO 18436-2 CVA Level 1
CAT I PREP I SERIES PRACTICE TESTS

Part	Covered Body of Knowledge	ISBN-13
1	PRINCIPLES OF VIBRATION	978-1-64415-006-1
2	DATA ACQUISITION	978-1-64415-009-2
3	SIGNAL PROCESSING	978-1-64415-002-3
4	CONDITION MONITORING	978-1-64415-005-4
5	FAULT ANALYSIS AND CORRECTION	978-1-64415-008-5
6	ROTATING EQUIPMENT	978-1-64415-014-6
7	ACCEPTANCE TESTING	978-1-64415-004-7
8	TWO PRACTICE TESTS	978-1-64415-007-8

Notes

Notes

Made in the USA
Monee, IL
05 November 2026

67816239R00118